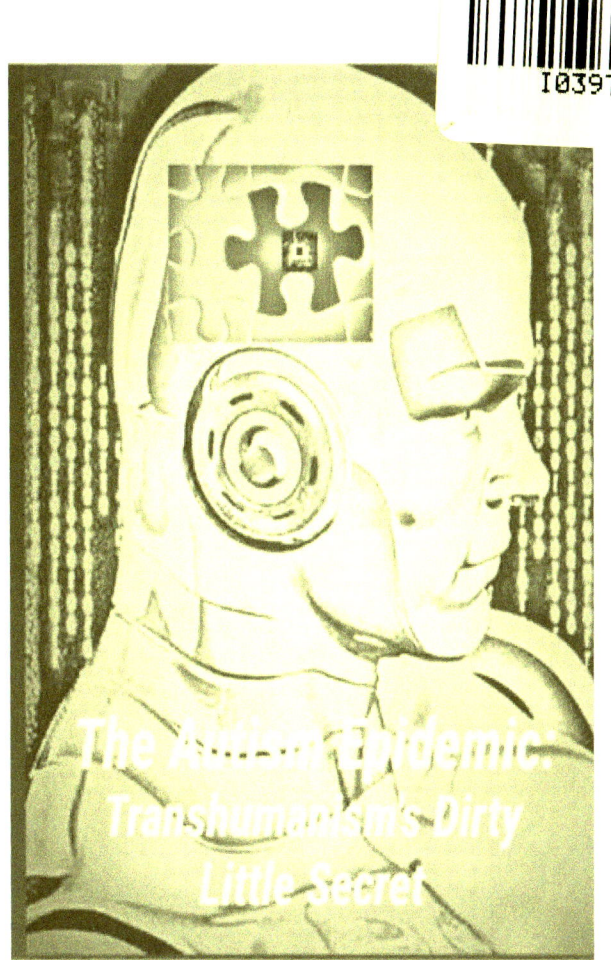

Wayne McRoy

The Autism Epidemic:
Transhumanism's Dirty Little Secret

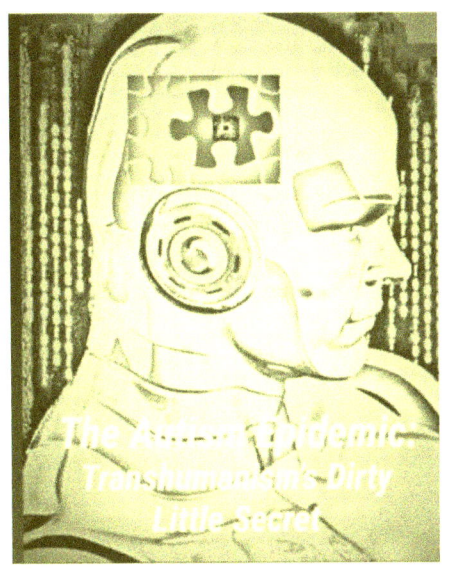

by Wayne McRoy

Copyright © 2019

All rights reserved

ISBN: 9781094678429

Contact Information:

I can be reached at:

alchemicaltechrevolution@gmail.com

youtube.com/Alchemical Tech Revolution

Or on Facebook at:

files from the conspiratorium

This book is dedicated to all who are affected by the autism epidemic, to all those that seek answers, and are not afraid to ask the difficult questions. Hopefully, together we can find answers to the origins of this epidemic, and make no mistake about it, it is an epidemic. Also, the potential exists to find solutions to the many problems associated with this condition. Let us not lose hope for our future progeny, for together, we can overcome...

I would like to thank my wife and children for inspiring me to write this book. Our struggles with the day to day rigors of autism have given me a unique perspective on the subject at hand. Two of our children are on the autism spectrum, so this subject matter is very personal to me. As I have firsthand experience in dealing with the ins and outs of this epidemic, I come from a very informed position in this matter. Many years of painstaking research has gone in to the volume before you. I sincerely hope that someone benefits from the information presented in this work...

What spirit dwells in the temple of transhumanism?

Know ye not that ye are the temple of God, and that the Spirit of God dwelleth in you? If any man defile the temple of God, him shall God destroy; for the temple of God is holy, which temple ye are.

1 Corinthians 3:16-17 KJV

The difference between
high-functioning and low-functioning
autism is that
high-functioning autism means
your deficits are ignored,
and low-functioning means
your assets are ignored

Table of Contents

Introduction - page 17

Chapter One - What is Transhumanism? - page 25

Chapter Two - What is Postgenderism? - page 33

Chapter Three - What is Autism? - page 43

Chapter Four - Autism and Gender Dysphoria - page 51

Chapter Five - Autism and Transhumanism - page 57

Chapter Six - Tavistock's Autism - Gender Dysphoria Link - page 65

Chapter Seven - The Neanderthal Theory of Autism - page 71

Chapter Eight - The Artificial Intelligence Factor - page 77

Chapter Nine - Cybernetics and the Foundation of a Control System - page 85

Chapter Ten - The Engineering of an Epidemic - page 95

Chapter Eleven - Artificial Intelligence and Re-engineering the Human Mind for Transhumanism - page 105

Chapter Twelve - Possible Causes of Autism and Methods of its Production - page 115

Chapter Thirteen - From Awareness to Acceptance - A Normalization Campaign - page 123

Chapter Fourteen - Rhetorical Autism and the Social Construct - page 131

Chapter Fifteen - Rewriting History to Support the Narrative - page 137

Chapter Sixteen - The Transhumanist/Enhancement Model of Disability/Impairment - page 147

Chapter Seventeen - Autism and the Military Industrial Complex? - page 157

Chapter Eighteen - The Origins of Autism - page 169

Chapter Nineteen - Nanotechnology, the Means to an End - page 179

Chapter Twenty - EMF Activation of Nanodevices and the Singularity - page 191

Chapter Twenty-One - Data Collection and the Blockchain Trap - page 199

Chapter Twenty-Two - Autism: The Alchemical Perspective - page 207

Chapter Twenty-Three - Conclusion - A Theory of Autism as an Engineered Epidemic - page 215

be why the mainstream media won't give equal time to stories about problems with vaccine safety.[15]

Conclusion

The safety of CDC's childhood vaccination schedule was never affirmed in clinical studies. Vaccines are administered to millions of infants every year, yet health authorities have no scientific data from synergistic toxicity studies on all combinations of vaccines that infants are likely to receive. National vaccination campaigns must be supported by scientific evidence. No child should be subjected to a health policy that is not based on sound scientific principles and, in fact, has been shown to be potentially dangerous.

Journal of American Physicians and Surgeons Volume 21 Number 2 Summer 2016

Introduction

In November of 2008, I began an unexpected journey into the world of autism. Our son was three months old, and had just been to the doctor for a routine checkup. He was a very happy, healthy baby, right on schedule with all of his developmental goals.

Like most parents, we listened to the doctor's advice, and allowed the clinicians to administer the recommended vaccinations for this visit. So, our perfectly healthy baby boy got his DTAP shot on that day, just as the vaccine schedule recommends. They told us he may run a fever and be a little fussy, that was perfectly normal.

Little did we know that he would have a severe reaction to that shot. Later that evening, our normally active, energetic baby was unusually lethargic. It was past his regular feeding time, so my wife tried to wake him up to eat. He was breathing very shallowly, and was unresponsive. He also had a grayish color tinge to his skin. We immediately ran him to the emergency room as quickly as we possibly could. We sat and watched helplessly in horror as our little angel that we had prayed diligently for for the past ten years lie dying in front of our eyes, as the staff in the emergency department seemed unconcerned for his well-being.

They seemed to think that we were just nervous first-time parents, who didn't know what we were talking about. This was, in fact, our second child, so that was not the case.

As we sat there in the waiting room for him to be seen, he made a very quick, remarkable recovery. By the time the doctor had seen him, he was awake, drinking his bottle, and looking much better.

The doctor examined him, found nothing wrong, and sent us on our way, with no plausible explanation for what had happened. Once again, I feel that the hospital staff dismissed the whole incident as nothing more than nervous, inexperienced first-time parents not knowing how to care for an infant.

Some days later, my wife was talking to a friend of ours who is a nurse. What had happened came up in conversation, and she told my wife that his reaction to the shot was a sign of autism, and she had seen it happen many times before. I didn't believe it at first, like most other people, I thought vaccines were harmless, and the only people who espoused the belief that they were harmful in some way were some Hollywood big shots that thought they knew more than the doctors. Hearing this from a nurse was a little unsettling. Our son seemed fine at the time, so we didn't put much stock into her observation.

A few weeks later, we started to notice something had changed. He started to drool excessively, and he didn't make eye contact with us anymore. He was also very colicky now, and he wasn't previously.

When we finally found out years later that our son had autism, we were devastated. We didn't know what to expect, or how to help him. He started exhibiting signs and symptoms at a very early age, and we were hard-pressed to find any doctors or medical professionals who were willing to help us. They kept telling us that he was too young to diagnose, come back when he starts school.

Unhappy with the lack of answers we were getting, I began doing deep research to find out what causes autism in an attempt to help our son have the most "normal" life possible. All we got from the medical community was the run around. No one had answers, no one was able to help us. We felt so alone and isolated. One of the comorbidities he suffers with his ASD is extreme ADHD, and this was the first and most difficult symptom for us to deal with at the time. We were exhausted. He was also quite the escape artist, so we had to constantly be vigilant. I began reading medical papers and case studies. I soon realized that there was something to the vaccine connection as well.

Eventually, I came to the realization that there was a synergistic combination of various environmental and genetic factors that correlated to autism and its symptoms. But, more importantly, I also came to the conclusion that there was a very real agenda behind the autism epidemic.

What I discovered in the course of my research and study, was far more shocking than the autism diagnosis itself...

What I discovered was behind the autism epidemic was an ideology called "transhumanism"...

"Be not deceived; God is not mocked: for whatever a man soweth, that shall he also reap."

- Galatians 6:7

Merging man with machine...

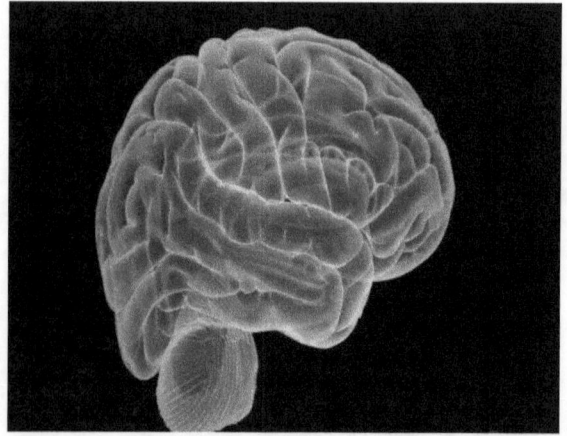

Chapter One - What is Transhumanism?

Transhumanism is a social and philosophical movement that is devoted to the research and development of "robust" human enhancement technologies[1]. These technologies consist of aspects of many different scientific disciplines, including biology, cybernetics, nanotechnology, genetics, and many, many more.

The term "transhumanism" was first used by Julian Huxley in 1957, in his paper of the same name, to describe what he considered to be the next step in human evolution. Since that time, the ideology behind the transhumanist movement has itself evolved from a philosophical idea into something more concrete, the merging of man with machine, the blending of the natural with the technological. This hybridization of the human form with advanced technologies is thought to be the catalyst of an evolutionary leap forward.

They refer to this "leap" as self-guided evolution. This is the transhumanist dream to become something much greater than man, something god-like, Human Plus. This evolutionary leap is the ultimate fulfillment of what the secret societies call "The Great Work". They also sometimes refer to it as "The Philosopher's Stone".

You see, the fact is the philosophy of transhumanism is far older than you are led to believe. In all actuality, its roots go way back to the ancient "mystery schools" of antiquity. These were the precursors of all of today's secret societies, such as the Freemasons, the Rosicrucians, The Order of the Eastern Star, and countless others whose uppermost levels collectively bind together in a separate order commonly referred to as "The Illuminati". It forms a secret society within a secret society.

Once you understand that the group behind the transhumanist movement is, in fact, this elitist "brotherhood" known as the "Illuminati", then you have to start to think about some key questions. What are they planning to do with transhumanism? Who will be the beneficiary? What are the spiritual implications? Who will be in control of this technology?

These are all serious considerations to keep in mind as we move forward towards what the transhumanists refer to as the "singularity". This is the point at which man and machine become one hybrid entity. Ray Kurzweil, one of the foremost personalities associated with the transhumanist movement (and a leading authority at Google), speaks quite frequently about this "singularity". One of his key works that outlines some of the plans of the transhumanist movement is a book called "The Age of Spiritual Machines".

As you can see, the people at the forefront of this agenda are not blind to the reality of the spiritual implications of transhumanism. To the contrary, they view the pseudo-spiritual aspect of transhumanism in high regard. There is actually an emerging "religious" movement surrounding the transhuman agenda.

Proponents of this pseudo-religious movement[2] believe that "the comfort provided by archaic religious superstitions impedes advancement and therefore should be set aside." and, "We might think of religion as premature transhumanism. Religion is not the opposite of transhumanism but a seed from which transhumanism can grow."

From what they write in their own publications, it would seem that the transhumanist movement has a very low opinion of traditional religious and spiritual beliefs. This philosophy promotes all of the worst aspects of modern "scientism".

The secular humanist belief system is the core philosophy at the center of transhumanism. The major players in this movement believe that there is no God, that mankind himself will be able to become God through this "self-guided evolution".

The advent of artificial intelligence is an integral part of the transhumanist revolution. With artificial intelligence as an extension of human consciousness, the proponents of this agenda believe we can achieve immortality. This will be accomplished by "uploading" our human consciousness into a computer. I know it sounds like science fiction, but plans for this possibility are already well underway[3].

Top transhumanist supporters claim that immortality will be obtainable by as early as 2029[4]. How will this be accomplished? Through the development of computer/brain interfaces, also sometimes referred to as "neuroprosthetics", man will soon be able to ostensibly upload their consciousness to the "cloud" for future download into a new biological and/or artificial body or modality. It is also feasible to transfer consciousness into a virtual reality world.

Transhumanism is far more than just augmenting your physical form with technological prosthetics, it is actually a vehicle for trying to understand the underlying secrets of "consciousness". Can this be accomplished and understood without the acknowledgement of a higher power, or the spiritual world?

Beneath the surface of the superficial appearance of this philosophy lies an ancient alchemical truth. This alchemical truth is being twisted and distorted by the modern transhumanist movement. This alchemical truth has to do with several principles collectively known as "Hermetic Philosophy".

So, in what ways does the transhumanist agenda distort these ancient hermetic principles? For an answer to this question, let's start first with the hermetic principle known as the "Principle of Gender". That is correct, the concepts of transhumanism and gender go hand in hand.

To better understand exactly what we mean by this statement, we will look at an agenda that runs concurrently with the transhumanist agenda, an agenda known as "Postgenderism".

These two terms, "Transhumanism" and "Postgenderism" can be thought of as being one and the same. Let's examine the "Postgenderism" philosophy and see exactly how it ties together with the agenda that is the transhumanist movement...

Baphomet -
The ultimate expression of Postgenderism

Chapter Two - What is Postgenderism?

In order to better understand the connection between the transhumanist movement and the postgender agenda, let's take a look at the definition of "postgenderism", as explained by a white paper from the Institute for Ethics and Emerging Technologies:

"Postgenderism is an extrapolation of ways that technology is eroding the biological, psychological and social role of gender, and an argument for why the erosion of binary gender will be liberatory. Postgenderists argue that gender is an arbitrary and unnecessary limitation on human potential, and foresee the elimination of involuntary biological and psychological gendering in the human species through the application of neurotechnology, biotechnology and reproductive technologies. Postgenderists contend that dyadic gender roles and sexual dimorphisms are generally to the detriment of individuals and society."[5]

As you can see, the postgenderist agenda calls for the use of technology to "liberate" people from the limitations of binary gender. Similarly, the transhumanist movement calls for the use of technology to help humanity transcend the limitations of its biological form. These two goals are intertwined, to end gender differences is to end a key characteristic of what defines what it is to be human. And, the end of humanity as we know it is the ultimate goal of the transhumanist movement.

Transhumanism and Postgenderism are synonymous with one another. You can't have one without the other. There are also two other interchangeable terms that describe this dualistic agenda to end gender and transcend the human condition. When you see these terms being used, you can know that they are clever codewords for "transhumanism".

The first of these terms is "Posthuman", and it means exactly the same thing as "transhuman". The proponents of the transhumanist agenda often use the term posthuman to describe what they consider the end result of this movement, something "beyond" human, the next evolutionary step of mankind towards "godhood".

The second term that is used interchangeably with "postgenderism" is a term that is very familiar right now in our society, and that term is "transgender". The proponents of the postgenderist movement see transgenderism as a transition phase from the gender binary to something superior, something better than the limitations of biological gender, a state of existence beyond male and female, a state beyond human.

The transgender movement is a gateway and a stepping stone to transhumanism. This is demonstrable from the writings of those who support, shape and promulgate this transgender ideology. It is a social engineering campaign to lead the unsuspecting masses straight into the transhumanist singularity.

For anyone who doubts that the ultimate goal of postgenderism aligns itself with the goal of transhumanism (that goal being the end of gender, and therefore also the end of humanity as we know it), here is a quote from a Master's degree thesis on the subject of postgenderism:

"The question is whether the borderline trans* identities might represent a departure to such a multitude existing under the heading of a less rigidly separated/ antithetical binary and whether the binary could be kept, yet radically weakened, in accord with the idea that 'the dance changes place and above all changes places. In its wake they can no longer be recognized.'"

In any case, this is an answer to the transhumanists and technofeminists who tend to think of gender inequality as something lodged in material circumstances of the body, easily done away by augmenting the body with technology. Technological, biological, etc., possibilities must always be discussed in the political frame of the present gender markers (or else they simply perpetuate misogyny): how much more probable is the idea that we may change their articulation than that we may erase them for good (or bad)? And how welcoming to difference is the idea that, to solve everything, it simply suffices we all follow in one direction?"[6]

The social engineering agenda surrounding this postgender/transhuman movement emerged shortly after World War 2. It was promulgated through the "New Age" movement, as well as the transpersonal psychology movement of the 1960s and 70s.

The New Age component of this social engineering campaign promoted the idea of "androgyny" as a form of spiritual enlightenment. This is, in all actuality, a perversion of the ancient alchemical teachings about the principle of gender.

In the psychological community, this agenda began to shift the view of the transgender individual from the concept of "gender dysphoria" to the reconceptualization of gender as being fluid, or a spectrum condition. Rather than treating it as a mental disorder, the psychological community began to lend credence to this condition.

These conditions set the stage for the emergence of a new, radical postgender subculture of transsexuals and transgendered individuals that began to grow and flourish. The ideology of a postgender society has now taken root.

The postgenderists soon realized that technology was their ultimate key to the society that they envisioned, and soon this ideology was merged with the transhumanist philosophy. The early studies in cybernetics helped to legitimize the potential of the transhumanist agenda. These two ideologies are having a very profound effect on our society at this current moment in history.

The combining of the postgenderist and transhumanist philosophies with the New Age Movement has created a massive spiritual vacuum that has sucked in the unsuspecting minds of millions of people hungry for a real spiritual walk. This manifests primarily as a form of secular humanism combined with a conception of moral relativism.

The coming pseudo-religion of transhumanism will be the ecumenical replacement of all other religious views if the controllers who pull the strings of society get their way. Becoming "Human Plus" will be the ultimate answer to the conundrum of gender identity as well as transcending the limitations of the human form.

So, now that we have established that the concurrent agendas of transhumanism and postgenderism are synonymous, the logical question you're probably asking yourself about now is "what does any of this have to do with autism?" The simple answer is "everything".

Before we progress any further, it will be necessary to define and identify what exactly autism is, and then we can begin to draw the lines that demonstrate that it is not just a random disorder that emerged by chance.

We will also refute the idea that autism is a condition that has been with us all throughout human history. In order to better understand how this all ties together, we will not only look at the medical condition itself, but we will also examine several "rhetorical" models of autism and their social engineering implications.

Likewise, we will examine multiple theories as to the origins of autism, and how they show definitive links to a much larger agenda, an agenda that leads directly to transhumanism.

Autism is much more than just a medical condition, as we shall demonstrate throughout this volume. There are several agendas at play that are directing the future of the autism epidemic.

This autism epidemic is a vehicle for something more, more than just a disparate, random set of symptoms, it is a blueprint for the destruction of the human "will", the means of absolute control of the mind. There is an entire social engineering campaign associated with it.

Let's start at the beginning, and first identify what the medical definition of autism is...

Putting the pieces of the puzzle together

Chapter Three - What is Autism?

Autism is a spectrum disorder that is characterized by challenges with social skills, repetitive behaviors, and issues with verbal and nonverbal communication.[7] Autism is also frequently accompanied by other medical issues, or "comorbidities", such as gastrointestinal conditions, seizures, sleep issues, ADHD, and other cognitive disorders.

Autism is considered to be a spectrum disorder. A spectrum disorder means that there are a broad range of symptoms that fall under the umbrella of the definition of the disorder, and every single case of the disorder is unique.

Each individual with autism exhibits different symptoms or combinations of symptoms.

Diagnosing ASD (Autism Spectrum Disorder) can be a difficult task. Diagnostic tools and procedures will vary from case to case, and although some cases of autism can be diagnosed as early as age 18 months, they more often than not take many years to arrive at a diagnosis. The most noticeable symptoms often have to do with a lack of understanding of social cues.

This also means treatment options will differ for each and every person on the spectrum. Some treatments work, or help, for some people but not others. Treatment options often include such things as behavioral analysis and therapy, medications, and adapting the patient's environment to better accommodate their individual challenges and needs.

What do we know about autism spectrum disorder? How many people suffer from this condition? Do we know what causes it? Does it affect certain groups of people more than others? How severe of a problem is this?

As of surveillance year 2006, it is estimated that 1 in 59 children have been identified with autism spectrum disorder.[8] This number only reflects those patients who have been diagnosed with autism spectrum disorder, which means that the numbers are likely much higher, as most children with autism are not actually diagnosed with the condition until they reach school age (usually around the ages of 4-5 years).

These numbers were compiled by the CDC in 2014, and have not since been updated. The numbers are likely high enough at this point to be considered an epidemic.

Boys are four times more likely to be diagnosed with autism than girls. ASD is reported to occur across all racial, ethnic, and socioeconomic groups. Almost half (44%) of children identified with ASD have average or above average intelligence and intellectual ability.

In the past, many of these children were diagnosed with a condition called "Asperger's Syndrome", which is now classified under the umbrella of Autism Spectrum Disorder.

For many years, conditions such as Asperger's Syndrome, PDD-NOS (Pervasive Development Disorder- Not Otherwise Specified), and various other conditions were diagnosed as a separate condition from autism spectrum disorder. Recent changes to the DSM (Diagnostic and Statistical Manual of Mental Disorders) now categorize all of these conditions under the banner of autism spectrum disorder. The DSM was updated to its fifth iteration in 2013 from the DSM-IV-TR to its current edition referred to as the DSM-5.

There is no known cure for autism. There are many treatments that can help alleviate symptoms. Some people have even claimed to "reverse" autism through special diet restrictions, chelation therapies, and homeopathic protocols. If these methods have truly worked for these people, that is fantastic, but no credible "peer reviewed" sources have ever been able to prove the credibility of these claims.

That's not to say that nobody has ever benefitted from these methods, it could just be that nobody has ever taken the time to verify that these methods work. Perhaps they do work for some people. Everyone's body is different, and reacts differently to various treatments. I do not personally hold out hope that the aforementioned methods hold the key to curing autism. For a spectrum disorder such as autism, I do not believe that there is a "one-size-fits-all" cure for it.

I believe that our best hope for eventually finding a cure for this disorder lies in identifying the underlying causes of it. Although the powers that be would like you to believe that autism has been with mankind all throughout history (there is a very real social engineering agenda that goes right along with this that we will discuss in a later chapter), we can demonstrate that it is only a more recent phenomenon.

Autism was not identified as an actual independent disorder until the 1930s. Starting with the work of Leo Kanner, and concurrently the work of Hans Asperger (the namesake of Asperger's Syndrome), the psychological community began to recognize autism as a unique mental disorder. Up until that time, they just identified it as a set of symptoms associated with schizophrenia. Leo Kanner is the one who put autism on the map.

What were Kanner's observations about the possible cause of autism? Dr. Kanner noted that the onset of the disorder, that later became known as autism, began following the administration of the small pox vaccine.[9] From the very beginning, vaccines have been implicated as a probable cause, but the pharmaceutical companies cover up and try to discredit this fact at every turn. It seems feasible that vaccines could possibly be a catalyst that triggers the onset of autism.

Now, having clearly defined what autism is, and some of the concerning facts about it, the question remains, "what does this have to do with postgenderism and/or transhumanism?" The answer is nothing short of shocking. Before we can get to this answer though, first we must do some dot-connecting.

How do these disparate agendas relate to the autism epidemic? The fact is that they are all interconnected in a very profound way. Let's begin our search for answers by drawing some direct lines from the autism epidemic to the postgender agenda...

Autism and Gender Dysphoria

Chapter Four - Autism and Gender Dysphoria

One of the prevailing theories that explains what autism is, is a concept called the "Hypermasculinity Model of Autism", also sometimes called the "Extreme Male Brain" theory of autism[10]. This theory postulates that gender differences can be classified into two different aspects of human thought: "empathizing" and "systemizing"[11].

Empathy is the ability to attribute mental states to other people, or the ability to understand that other people have their own thoughts, emotions, and desires. Empathizing with someone means you understand what they are thinking or feeling. The attribute of empathy is widely regarded to be a more commonly feminine trait.

In contrast, systemizing is defined as a tendency to organize things and develop principles to understand complex systems. This attribute is commonly regarded to be a more masculine trait. Studies have shown that males tend to have more of a propensity to systemize, whereas females have a higher propensity to empathize.

Studies have been done comparing these traits in individuals with ASD to control groups of neurotypical children, and the outcomes support the "extreme male brain" theory of autism. The individuals with autism scored consistently higher on systemizing tests than the neurotypical children, but likewise consistently scored lower on empathizing tests than the neurotypical control group.

Although this theory is somewhat controversial, there is a plethora of data that supports its merit. One study has shown that children who were exposed to higher levels of testosterone in the womb have a greater chance of developing autism-associated traits, regardless of gender.

This study suggests that these higher levels of fetal testosterone exposure have a tendency to "masculanize the mind", lending credence to the "Extreme Male Brain" theory[12]. It would appear that testosterone plays a part in autism spectrum disorder. Could this somehow relate to some sort of genetic predisposition for higher fetal testosterone levels?

There is another theory of autism that relates to testosterone via a genetic predisposition, and likewise, also suggests that genes related to autism play a part in the evolution of intelligence. This theory is called the "Neanderthal Theory of Autism". Researchers claim that a "residual echo" of Neanderthal DNA[13] affects portions of the brain that have been correlated to autistic behaviors.

The Neanderthal theory postulates that certain intelligence factors, such as visual and spatial acuity are enhanced in a sort of trade-off with other factors related to social intelligence in individuals with autism. It further postulates that this neurological diversity played an important role in the evolutionary development of modern man[14].

So, where does testosterone fit into this picture? Genetic evidence would seem to suggest that a large drop in overall testosterone levels occurred starting about 200,000 years ago (according to the "evolutionary model"), seemingly as a result of pre-modern humans interbreeding with Neanderthals[15].

Supposedly, the interbreeding of these pre-modern human species led to the rise of modern man, who, due to the drop in testosterone levels, developed greater social skills and communication techniques, allowing them to become the dominant species. But, genetic markers of these pre-modern human species are still passed on as recessive traits even today. Indeed, Neanderthal genes have been identified and tracked in scientific studies.

These Neanderthal genes have been linked to different facets of human intelligence, mostly relating to spatial and/or visual intelligence[16]. It would also appear that people with these recessive Neanderthal genes tend to have higher than average testosterone levels, as well as a higher prevalence for autism, which may correlate back to the "Extreme Male Brain" theory.

Could there be something to this? We've been able to show that there is a potential link between Neanderthal genes and high testosterone. Does this testosterone link in the autism epidemic somehow relate to transgenderism[17]?

Recently, the Tavistock Centre was forced to admit a high prevalence of autistic clients at its clinic for transgender children[18]. Nearly one third of all of the children being treated at the clinic had some form of autism. Does this indicate that the Tavistock Centre was taking advantage of youths with clear mental health challenges, who were experiencing some confusion that the chaos of day to day life can bring to people on the spectrum? Or, is there a legitimate link between autism and transgenderism?

Either way, there is clearly a social engineering agenda at work trying to steer transgender ideology and autism in the same direction. The question then becomes, why? What is the purpose of trying to "normalize" the autism epidemic and transgenderism, which, consequently, used to be classified as a mental disorder known as "Gender Dysphoria"?

This question leads us to our next step in deciphering the enigma of the autism epidemic. We have now drawn the lines showing the link between autism and transgender ideology, which pushes these two distinct disorders (autism and gender dysphoria) in the same direction. And that direction is straight into the Postgenderism agenda, which lines up directly with the Transhumanist agenda...

Chapter Five - Autism and Transhumanism

How is the autism epidemic being used as a vehicle to transhumanism? Why is the autism epidemic being used as a vehicle to transhumanism? Why are things like gender identity and Neanderthal genes so important to understanding the agenda that is known as transhumanism?

In order to better understand the inner workings of the transhumanist agenda and its related social engineering campaign to normalize and assimilate autism and gender dysphoria into its coming "singularity", we will first firmly establish the fact that there is a definitive link binding together autism, gender dysphoria, and artificial intelligence.

That is correct, you read that right, artificial intelligence plays a major role in this agenda. As profound and astonishing as it may seem, the autism epidemic has everything to do with artificial intelligence. Before we elaborate on this point any further, let's first connect some more dots to establish the trail that leads down the transhumanist path.

This all has to do with garnering a basic understanding of how intelligence and consciousness work. This interdisciplinary field is known as "Cybernetics", which we will discuss in detail in a later chapter. The rise of cybernetics[19] in the 1930's and 1940's led to advances in the research and development of cognitive sciences and behavioral analysis, as well as technological breakthroughs in fields like artificial intelligence.

Cybernetic models of intelligence and consciousness began to be studied and theorized around the same time that autism became a separate diagnosis from schizophrenia. This is not a coincidence, and we will cover this more later when we take a more in-depth look at the history of cybernetics, and how it affected modern psychology and physiology.

Autism is a condition that isolates out certain facets of intelligence, which makes it easier to study these particular facets of intelligence. Similarly, gender differences complicate the study of these particular facets of intelligence, so it is desirable for those who study behavior to have a subject that is less affected by classical gender differences.

This makes autistic individuals desirable test subjects for understanding the nature of intelligence and consciousness, without all of the emotional attachments that normally make analysis of critical information very difficult. This is also the reasoning behind the push for postgenderism, removing gender, and all of the emotional baggage that comes with it, from the equation makes the study of consciousness and intelligence easier to quantify.

That is the bottom line to all of this, the powers that be are trying to quantify intelligence and consciousness. In my view, intelligence and consciousness are inseparable from one another. Quantifying intelligence and consciousness makes it controllable and transferable. This concept will be important later when we discuss computer-brain interfaces.

Understanding how the mind works will allow scientists to create artificial intelligences that may or may not be considered conscious beings. Going one step further, the ability to create artificial intelligences that have conscious awareness is the key to transhuman enhancement, the merging of the minds of men with these artificial intelligences.

Understanding the autistic mind is pertinent to the development of artificial intelligence. Scientific studies have shown that the autistic brain works very much like a computer. In fact, IBM's artificially intelligent machine known as "Watson" has been compared to an autistic brain by the very scientists who helped to develop it[20].

So, let's take a step back for a moment and review what we already know. We have been able to establish that there is a more than coincidental correlation between autism and gender dysphoria. We have also shown that there is strong evidence that suggests that autism is intrinsically linked to Neanderthal genes. We have also demonstrated that the autistic mind is the key to developing artificial intelligence.

So, can we show that these seemingly disparate bits of information have some sort of merit to them? Are there studies being conducted that compare autism and artificial intelligence? If autistic intelligence is so important to artificial intelligence, is there any research to back up that link?

Likewise, if autism can be traced back to Neanderthal DNA, are there any studies showing that link, and if so, would that not mean, by proxy, that Neanderthal intelligence is also important to the development of artificial intelligence? Are there any studies or experiments being done that link artificial intelligence to the Neanderthal brain?

Let's begin first by looking at some studies showing the strong prevalence of gender dysphoria among those with autism. Second, we will address studies displaying the autism-artificial intelligence link, and lastly we will analyze studies indicating the Neanderthal link to autism and discuss the ongoing experiments wherein scientists are growing Neanderthal brains to power and animate robot bodies[21]. Yes, you read that right.

To better understand the connection between autism and gender dysphoria, let's take a look at the Tavistock Centre...

The Autism Epidemic: Transhumanism's Dirty Little Secret

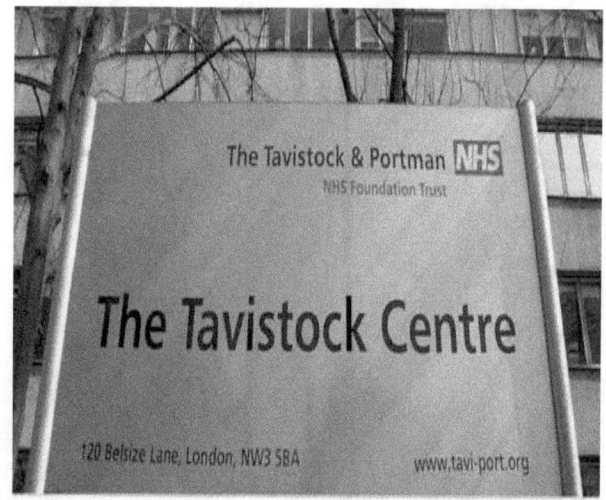

Chapter Six - Tavistock's Autism - Gender Dysphoria Link

Few institutions in this world have had a more far-reaching effect on society than the Tavistock Institute. Tavistock has been implicated in many diverse psychological manipulations and social engineering campaigns through the years since its inception. This one organization has had its fingerprints all over international agendas and projects, especially agendas that affect group behavior and population policies.

A subsidiary of the Tavistock Institute, the Tavistock Centre (also known as the Tavistock and Portman NHS Trust) began its Gender Identity Development Services program in 1989, and later opened Britain's first transgender clinic for children. This clinic has recently been criticized for covering up important data that indicates that more than one third of the patients treated at the clinic between 2011 and 2017 displayed moderate to severe autistic traits.

What does this information mean? Is there a substantial link between autism and gender dysphoria? Or, is there an agenda at play at Tavistock to push transgenderism onto children with other obvious mental health issues? Or, perhaps, could there be a little bit of both of these possibilities going on[22]?

There have been numerous other studies that show a high prevalence rate of gender dysphoria within the autism community. A 2017 article in Forbes magazine lists links to nine separate studies that indicate a correlation between the two[23].

Comorbidities with autism are not uncommon. Perhaps gender dysphoria can be considered a comorbidity with autism.

Of course, that brings up another question. Is gender dysphoria considered a mental disorder? There is a large gray area surrounding this subject. Currently, there is a "normalization campaign" underway to make transgenderism socially acceptable. Although technically considered a mental health disorder, it is largely accepted by the psychiatric community as a relatively "normal" biological condition.

Although gender dysphoria (which used to be called Gender Identity Disorder) is a condition outlined in the DSM-5 (this is the Diagnostic Statistics Manual, the book that categorizes and identifies all known mental health disorders), it is generally not treated as a mental health disorder anymore. In fact the name was changed to eliminate the stigma of being referred to as a disorder.

It could legitimately be argued that the issue of gender identity has become so politicized to the point that people with an actual psychological condition often do not seek or receive treatment for their disorder, with the exception of hormone therapy if they decide to transition to the gender that they "identify" with.

To take a common sense stance on this issue and identify people by their biological gender is no longer socially acceptable.

This is part and parcel to how social engineering works. Societal norms and popular opinion make you fearful to tell it like it is, instead you have to be extremely careful of the words you choose and the way you describe things so as not to offend anyone. This is an affront to free speech, and a disservice to those who suffer from these psychological disorders, all for the sake of a political agenda designed to separate us, in order to make us all easier to control.

Could there be a common cause for both autism and gender dysphoria? It seems highly unlikely that these two conditions would overlap in so many ways if there was not some kind of common source or causality. Perhaps this would lend some credence to the "Extreme Male Brain" theory of autism, as it has been evidenced that high testosterone levels are a more common theme among those on the spectrum[24].

So, although there is a notable prevalence of gender dysphoria associated with autism, could it be possible that maybe there is some kind of incentive for Tavistock's Gender Identity Development Services program to steer patients into therapeutic treatment for gender confusion? Is there a large sum of money tied to this program?

These two questions go beyond the scope and intent of the subject of this volume, so we will leave these speculations open-ended for now. These two questions did need to be addressed, however, because it would be irresponsible to ignore the fact that 372 out of 1069 patients being treated for gender identity issues at the Tavistock Centre were identified as having moderate to severe autistic symptoms, and the Centre tried to hide the data.

So, now we've looked at the autism-gender dysphoria link, let's move a step further along the transhuman road to the Neanderthal theory of autism...

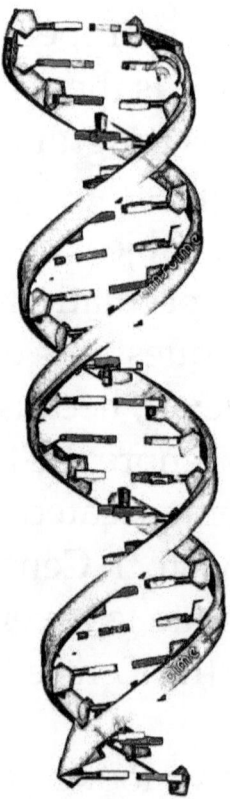

Chapter Seven - The Neanderthal Theory of Autism

One of the more intriguing theories of autism is the Neanderthal Theory of Autism. We have already established that there would seem to be a potential correlation between Neanderthal genes and autism, as well as a link to fetal testosterone levels. The Neanderthal theory introduces an evolutionary component to the mix.

With these interesting connections in mind, the Neanderthal theory tends to validate the hypermasculinity, or extreme male brain model of autism, a well as autism's track record with gender dysphoria. These seemingly unrelated conditions do overlap with one another, as borne out by scientific studies and corroborating evidence.

So, what do we know about the Neanderthal Theory of Autism? Neanderthal theory was first suggested by a researcher named Leif Ekblad in 2001. The theory postulates that hybridization between Neanderthals and modern humans in Eurasia led to neurodiversity as an adaptation for species survival, and genetic traces of this interbreeding are evident today, particularly in people with autism and other related neurological disorders[25].

Since the theory's inception, other scientific disciplines have produced supporting evidence for this hypothesis. For example, archaeological studies claim that evidence suggests that the rise in "collaborative morality" in early humans led to the acceptance of neurodiversity, thus contributing to the survival and evolution of the species[26]. These findings claim that conditions like autism provided specialized skills[27] that helped early man to adapt and survive, and that this rise in collaborative morality accorded people with these conditions a certain respect.

Additionally, genetic research supports the argument that "autism without intellectual impairment" became a significant adaptation that helped early man survive and thrive[28]. All of these interdisciplinary studies tied together demonstrate evidence to support the Neanderthal theory.

The Neanderthal theory further states that neurodiversity (i.e. - conditions like autism, schizophrenia, ADHD, etc.) is a "fully functional human variation" related to typical adaptations expected in a species. This line of reasoning has steered the impression of autism in the direction of being the key to the next step in human evolution.

This next step in human evolution is intrinsically linked to the development of artificial intelligence. Comparisons between artificial intelligence and autism are often cited in artificial intelligence research and development. The connection between autism and artificial intelligence was initially established by consciousness and intelligence studies conducted by the cybernetics group established with funding from the Macy Foundation and Rockefeller Foundation.

Cybernetics is the key to understanding the rise in the autism epidemic, and is also the philosophy that birthed the modern transhumanist movement. An examination of the history of cybernetics and its studies into consciousness and intelligence is imperative to getting to the bottom of the conundrum that faces us today.

But, before we dive deep into the rabbit hole to take an in-depth look at the role of cybernetics in all of this, let's first establish how artificial intelligence interconnects with autism. We will also examine how the Neanderthal link also ties to artificial intelligence, as this all combines together in very profound ways.

Autism, consciousness studies, Neanderthal robots, transhumanism, mental disorders, postgenderism, and social engineering, this all sounds like a bad science fiction movie. But, sadly, it's all true, and verifiable through mainline scientific research by anyone who is willing to take the time to look for it. The philosophical, spiritual, and physical ramifications of the transhumanist agenda are profound.

Let's take our next step down the transhumanist path and examine the wonders of artificial intelligence...

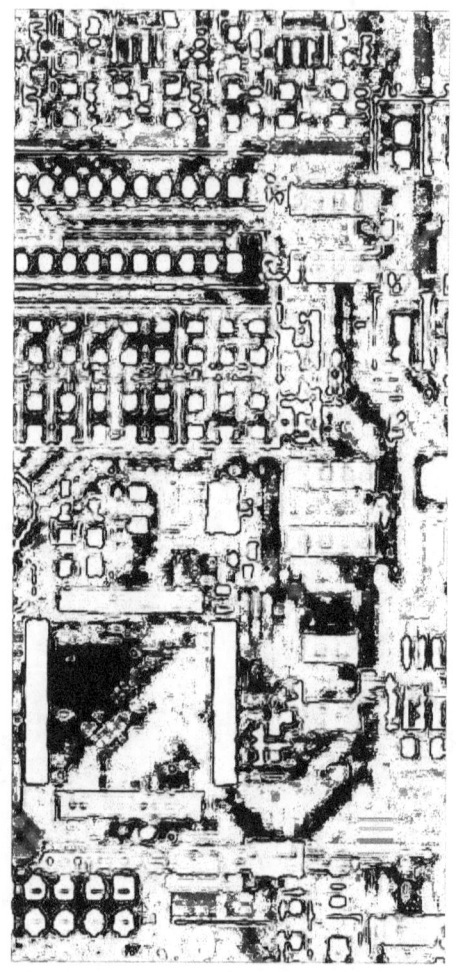

Chapter Eight - The Artificial Intelligence Factor

Perhaps the most important aspect of the transhumanist agenda is the development of artificial intelligence. Although the idea of artificial intelligence can be traced back to the myths and legends of antiquity, the modern science we associate with the subject found its origins in the work of the early cyberneticists. It is through the cybernetics approach that the advent of artificial intelligence has become an achievable goal. We will look at this in more depth later.

The question we need to consider now is why is artificial intelligence so important to the transhumanist agenda? And, what does any of this have to do with autism?

Let's take a look at some comparative studies that show that autism learning has significant correlations to artificial intelligence learning. This is one of the key points in understanding how to develop what is called general artificial intelligence.

One important aspect of autism is a focus on visual learning and spatial intelligence[29]. This relates to a model of intelligence called the VPR model of intelligence[30]. VPR stands for visual, perceptual, and image rotational. This is a model of general intelligence that explains much of what we know about autism learning, and is one of the most compatible models of intelligence to use for the development of a general artificial intelligence.

This model of intelligence incorporates two distinguishable facets of intelligence that help to explain why those on the autism spectrum have certain strengths and weaknesses in their learning abilities. These two facets are called fluid intelligence and crystallized intelligence.

Fluid intelligence can be best described as entrenched capacities associated with reasoning. This represents a person's innate abilities. By contrast, crystallized intelligence is defined as things learned over time. This would be things like vocabulary, skills, and mathematical concepts. These are learned capacities.

People on the autism spectrum tend to test high in fluid intelligence tests, but often lower in crystallized intelligence tests. This dichotomy relates to a concept called neuronal plasticity. Neuronal plasticity primarily reflects processing efficiency.

Those on the autism spectrum often seem to display deficits in processing efficiency due to the disregulation between their capacities of fluid intelligence and crystallized intelligence. So, although they process information at a higher rate of speed than neurotypical, this disregulation translates as an inability to read social cues and understand certain subtleties of language, such as humor, wordplay, or sarcasm.

This, however, does not mean that those with autism are not intelligent, in fact, the evidence would indicate that the opposite is true, that autism is a disorder of high intelligence. Autistic people tend to have a higher level of perceptual intelligence than neurotypicals. That is where this VPR model of intelligence comes into play. These people display greater visual and spatial acuity than people who are considered neurotypical.

This higher visual and perceptual intelligence has been genetically attributed to Neanderthal DNA[31]. And, once again, we can see where that model of autism supports what we can demonstrate to be true. So, this VPR model of intelligence is one of the primary models used to understand and manipulate the study of consciousness systems, especially in non-neurotypical people.

Computers operate under similar parameters. They can be likened to people with disregulated balances between fluid and crystallized intelligence. Machine intelligence lacks the same degree of neuroplasticity as humans, so this translates similarly to corresponding inabilities within autism. For example, computers read language literally, and may not understand metaphors (in much the same way as people with autism). The computers' inability to understand these types of subtleties make development of artificial intelligence difficult.

However, by understanding how autistic learning works, those who study and develop artificial neural networks have been able to produce some great leaps forward in regard to general artificial intelligence. By designing the artificial intelligence systems to mimic the disregulated VPR model typically seen in people with autism, scientists have been able to produce some remarkable results with these neural networks. They've even been able to produce machines that can pass the Turing test[32].

So, understanding that the use of autism as a model to produce general artificial intelligence can help bring about the development of a greater than human intelligence neural network system, we can begin to see why this autism epidemic is such a key tool to the goal of the transhumanist singularity. And that brings us to our next point, if the autistic brain is more compatible with artificial intelligence, then it would stand to reason that the Neanderthal brain would likewise be more compatible to artificial intelligence.

This would explain why there are multiple experiments that are growing Neanderthal neural cells on computer chips[33], with the eventual goal of having these "neanderoid brains" powering and operating robots. This is not the stuff of science fiction, these experiments are occurring as we speak. There have been several recent news articles talking about this.

So, we can see that a big part of the overall agenda is the development of artificial intelligence systems that are compatible with the autistic brain. The next step in the transhumanist plan will be to connect artificial neural networks to human brains through a device called a neuroprosthetic.

This is a computer-brain interface device. One that has been talked about a lot lately is something called neural lace, a product being developed by companies such as Elon Musk's Neuralink, or a lesser known company called Kernel (and this is a company to keep an eye on, there is reason to believe that their research and development may be further along than Neuralink's).

To better understand how the advent of artificial intelligence corresponds with the escalation in the autism epidemic, let's take a look back in history to the modern start up phase of the artificial intelligence field, and where it overlaps with the "discovery" of this condition that we call autism.

To find this start point, we must now thoroughly examine the scientific interdisciplinary philosophy known as cybernetics, and this is the key to unlocking the answers we seek about the how and why of the autism epidemic...

The Macy Conferences

Chapter Nine - Cybernetics and the Foundation of a Control System

Cybernetics can best be defined as an interdisciplinary holistic philosophical approach to studying, manipulating, and controlling systems. It was born out of World War 2's "operations research" used for logistics purposes to aid the allied campaign against the axis powers.

Cybernetics utilizes a holistic approach to whatever is being studied. This means that the cyberneticists study things from the vantage point of whole systems rather than individual constituent parts. Much practical information can be gathered, and much critical control can be achieved through the analysis of whole systems as compared to separate individual facets of a system.

An initial meeting in 1942 between several people from diverse backgrounds and scientific disciplines led to the creation of the "cybernetics group". This group officially came together en force in 1946, after the second world war ended, and spawned the now famous Macy Conferences[34].

The Macy Foundation, along with the Rockefeller Foundation, provided the early funding for the cybernetics group, and the Macy Foundation also hosted and promoted the conferences. They saw great promise in the work of the cybernetics group, and provided them with the resources necessary for them to pursue their work.

Who were the major players in this cybernetics group? What types of things were they studying? Who was pulling the strings behind the scenes? And how does all this relate to autism? We will explore all of these questions, and more, in order to understand how we got to where we are today, and where all this will be going in the future.

The original cybernetics group consisted of the following movers and shakers in their various fields of study, motivated and put together by the Macy Foundation's Lawrence K. Frank and Frank Fremont-Smith. This is just a partial list of some of the more influential members:

Warren McCulloch
Walter Pitts
Claude Shannon
Donald MacKay
Harold Alexander Abramson
William Ross Ashby
Yehoshua Bar-Hillel
Gregory Bateson
Margaret Mead
F.S.C. Northrop
John von Neumann
Norbert Wiener
Paul Lazarsfeld
Kurt Lewin
Heinz von Foerster
Julian H. Bigelow
Arturo Rosenblueth

This is an impressive list of recognizable names. These were people of accomplishment. And they all played a part in the development of the study of cybernetics.

Before we start connecting some dots, let's clarify what cybernetics is and where it comes from. The term cybernetics is derived from the Greek word "kybernetes", which means "pilot" or "steersman". So, cybernetics is about control. The term cybernetics was coined in 1948 by mathematician Norbert Wiener to describe this new science of systems control.

Cybernetics has been described as "the science of effective organization", and as "the science of control and communication in the animal and the machine".

A few key aspects to a better understanding of cybernetics are the concepts of "circularity" and "homeostasis".

Circularity has to do with information input and output flows in a closed system. Alan Scrivener explains the idea of circularity in cybernetic study as follows: "Cybernetics is the study of systems which can be mapped using loops (or more complicated looping structures) in the network defining the flow of information. Systems of automatic control will of necessity use at least one loop of information flow providing feedback." So, to simplify, cybernetics uses feedback loops as a methodology of control of whole systems.

Another important component of the cybernetic approach to understand is the concept of "homeostasis", especially as it relates to biological systems. Two of the key proponents for the concept of homeostasis were senior executive of the Macy Foundation, Lawrence K. Frank, and his protege, Frank Fremont-Smith, head of the Macy Foundation's medical office. Frank and Fremont-Smith both had longtime interests in child development.

Frank was intrigued by Walter Cannon's 1929 work on homeostasis, and how the concept might pertain to child development. Fremont-Smith helped establish a network on the subjects of neurophysiology and homeostasis, and how they interrelate. Thus, the homeostasis idea became one of the key foundations of cybernetics.

But, what exactly is homeostasis? Homeostasis[35] is based upon a concept first introduced by French experimentalist Claude Bernard in the mid-nineteenth century. Bernard called this concept "milieu interieur", meaning "inner world". This idea was later expounded upon by Cannon, and renamed "homeostasis".

According to Cannon's homeostatic model[36], the brain coordinates body systems, with the goal of maintaining regular internal variables within a constant range. It proposes that the body (and mind) are self-sustaining and self-regulating machine systems that constantly strive to maintain internal balance.

This theory was later seized upon by the cyberneticists to help figure out how to quantify and control biology and consciousness systems. And here is where the proverbial rubber meets the road, the real meat and potatoes of the subject.

From cybernetics are born the concepts of artificial intelligence and transhumanism. Both of these are outgrowths from the scientific endeavor to quantify and control biology and consciousness. This is also the fallow ground for the birth of the autism epidemic, as we will discuss further.

In order to better understand how these seemingly unrelated topics join together, let's look at a little bit of the history of the developing fields of psychology, sociology, and child psychology from the vantage point of the cybernetics approach, and its relationship to the Macy Foundation (and the people pulling the strings from "behind the veil").

These scientific disciplines were created for the purpose of quantifying and controlling consciousness, using the whole systems principles implemented by the cyberneticists. And, the agenda, and the push of the cybernetics movement, was controlled from the top down by the same small groups of financiers supplying the resources to the scientific community, especially those working with the cybernetics approach.

Through these methods, the controllers behind the scenes, and the financial backers of these scientific studies, acquire access to all the data produced by the scientists whom they funded. Through the selective use of this data, agendas are driven and steered to make certain scientific developments and breakthroughs by the power structure. This methodology of control by the few at the top of the power pyramid continues even today. There is an agenda that was born out of cybernetics that continues right up to the present time.

But, before we look at that agenda, let's examine two very key components to what we are analyzing here today. Looking at these components will answer the question, what does this have to do with autism?

We are going to look at the role of one of the most influential players in the science of cybernetics, Gregory Bateson. His "Theory of Mind" and a collaboration he put together called the "Bateson Project" are the foundation from which the autism epidemic springs forth...

Gregory Bateson

Chapter Ten - The Engineering of an Epidemic

One of the key foundations for understanding the processes of mind and consciousness lies within the work of Gregory Bateson. Bateson's use of cybernetics methodologies to map out consciousness and neurological systems created a start point for later cyberneticists to build upon in order to create a functional model of how human intelligence works from a whole systems perspective, and to later figure out how to manipulate human intelligence for the purpose of control.

Remember, control is the primary purpose of cybernetics. Creating a system of control for the entire domain of human intelligence has been a long sought after goal by those in positions of power.

Indeed, it would appear that a shadowy "elite" have seized upon the opportunity to hijack the work of the early cyberneticists, who sought to understand the workings of the mind, and twisted it into a means to construct a control system to force humanity into a false "evolutionary leap".

Before we examine this avenue of thought, let's take a look at the work of Gregory Bateson in order to understand how certain elitist interests took hold of the cybernetics approach and used it to try to remanufacture human consciousness into a type of artificial intelligence.

Bateson initiated a collaborative program called the Bateson Project[37], a comprehensive study into schizophrenia and related mental disorders, in order to understand how intelligence and consciousness systems work under normal circumstances, as compared to what happens in regards to systemic homeostasis in patients with these psychological conditions.

The project originally tried to link the behavioral disturbances associated with schizophrenia to external non-physiological factors. Bateson came up with a theory called "double bind" which sought to explain this behavior[38]. After further study, it was concluded that, although the double bind concept had some merit to it, ultimately, there had to be a physiological factor involved that caused schizophrenia.

Keep in mind, at this point in history, autism had not caught on as a diagnosis yet, so, patients with autism were lumped in with and categorized as schizophrenics, or pre-schizophrenics in the case of children. Indeed, there were children involved in the Bateson Project.

The idea of autism as a separate condition from schizophrenia came about from the work of Leo Kanner[39]. Kanner was recruited and funded by the Macy Foundation and Rockefeller Foundation to set up a small pediatric clinic where he later did his now famous studies on autism[40].

As we mentioned earlier, the Macy Foundation would most definitely have had access to Kanner's data. The Macy Foundation also had a vested interest in the cybernetics studies that were being performed by Bateson and company. It is probable that Bateson was familiar with Kanner's work. What kinds of important scientific breakthroughs came about from the combination of Kanner's discoveries and these cybernetics studies?

Bateson layed down some very important foundational theories of how the mind works. Bateson's model of the theory of mind[41] outlines some important distinctions that will be important to understand the mechanisms used for the purpose of control.

Let's start by outlining what Bateson's definition of mind is.

The following precepts are Bateson's "Criteria of Mind" as conveyed in his 1972 book titled "Steps To An Ecology Of Mind":

1) Mind is an aggregate of interacting parts or components.

2) The interaction between parts of mind is triggered by difference.

3) Mental process requires collateral energy.

4) Mental process requires circular (or more complex) chains of determination.

5) In mental process, the effects of difference are regarded as transforms (that is, coded versions) of the difference which preceded them.

6) The description and classification of these processes of transformation discloses a hierarchy of logical types immanent in the phenomena.

So, Bateson defined mind as a sort of machine-like system (this is a closed system) that functions based upon a hierarchy of inputs (or you can think of these as symbols or signals) which link together by a cause and effect energy transfer, resulting in a complex branching interconnection of chains of causation. These interconnections create a causal circuit, the basis for controlling the system. This chain of causation can be traced all around the circuit and back to whatever position was the starting point for action. In such a circuit, events at any position in the circuit have effects at all positions on the circuit at later times.

The causal circuit conveys this information/energy change through the entire system through a process called "feedback". When an external variable disrupts the circuit at any point in the circuit, it generates a whole system response that repeats through the system until the variable is resolved. This results in a "feedback loop".

Bateson says, "In principle, then, a causal circuit will generate a non-random response to a random event at the position in the circuit at which the random event occurred."

This causal circuit is used to change the value of a variable at a later time when the sequence of effects has come around the circuit. This concept can be used to initiate a certain effect at a specific part of the circuit by choosing the right area as the start point for the chain of causation.

So, what does this all mean exactly? It means that it is possible to manipulate the mind system by creating a controlled, repeatable effect at a specific start point within the causal circuit, and taking advantage of the homeostatic mechanism that regulates the circuit. So, in theory, using the correct type of stimulus or substance to trigger a desired reaction of the body's or mind's homeostatic defense mechanism could be used to produce a specific outcome.

For example, it could be speculated that this concept can be used to target a specific area of the brain with a specific stimulus or substance (we'll say in our hypothetical example here that perhaps we'll use aluminum nanoparticles) in order to induce the homeostatic defense mechanism to produce a certain effect (in our hypothetical example, we'll say we are trying to produce autism symptoms) by using the nervous system as a causal circuit to create a feedback loop.

So, now that we understand the underlying concept of how this works from a scientific cybernetics perspective, we can examine what the real world effects of something like this would look like to the external observer.

Disregulation of neuronal pathways causes a system imbalance, resulting in a homeostatic feedback loop in patients with autism, schizophrenia, bipolar disorder, and various other similar psychological disorders. This disregulated feedback loop appears outwardly as things like repetitive behaviors, difficulty with communication and social interaction, obsessive compulsions, inappropriate anxiety, delusions, and lack of empathy, among other symptoms.

It has now been demonstrated that the cyberneticists figured out a general idea of the mechanisms that caused these psychological disorders. It has also been shown that these conditions could be reproduced in a laboratory setting[42]. The next logical step in the process would be to reverse engineer these conditions to figure out solutions to these problems. What happened from there was the birth of artificial intelligence studies.

The study of artificial intelligence is critical to understanding how the human mind works. The development of artificial intelligence arose as a direct result of the attempt to reverse engineer human consciousness as it came to be understood through the study of these homeostatic disregulations present in these patients with schizophrenia and autism. Let's look at the rise of artificial intelligence...

Artificial Intelligence

Chapter Eleven - Artificial Intelligence and Re-engineering the Human Mind for Transhumanism

The quest to develop artificial intelligence began in earnest in 1950 with the introduction of Alan Turing's paper, "Computing Machinery and Intelligence". Based upon Turing's concepts, five years later, Allen Newell, Cliff Shaw, and Herbert Simon introduced a program called "Logic Theorist". This program was designed to mimic the problem solving skills of a human being.

"Logic Theorist" was funded by the RAND Corporation. This was the first acknowledged artificial intelligence program.

"Logic Theorist" was presented at a 1956 conference organized by John McCarthy called the Dartmouth Summer Research Project on Artificial Intelligence[43].

It was John McCarthy who first coined the phrase "artificial intelligence". This conference launched the developmental field of A.I. research.

Soon after, proponents of the emerging field successfully advocated for funding from government and quasi-government agencies, most notably DARPA (Defense Advanced Research Projects Agency) and IARPA (Intelligence Advanced Research Projects Agency). Over the next twenty years, artificial intelligence research was very slow, as the computing power was still just too weak to exhibit intelligence.

Funding for A.I. projects dwindled until some important breakthroughs were made in the 1980s. New algorithms made possible the concept of "deep learning" techniques. Computing power improved also.

By the 1990s and early 2000s, many of the landmark goals of artificial intelligence research had been achieved. The fundamental problem of computer storage capacity that had held back development in the artificial intelligence field for nearly thirty years was no longer an issue. As a result, we now live in the era of "big data".

The A.I. field is currently actively working on the long term goal of developing a general artificial intelligence machine[44] that can bypass all human cognitive abilities in all tasks. But, despite some major advances in modern technology, neural networks haven't changed much in forty years. This presents some problems that need to be overcome.

The human brain is far more efficient than neural networks. In neurons in the brain, only 5% to 10% of the input is actually coming from the previous layers. Information is instead garnered from other neurons "downstream", so the information circuit has a lot of loops and is more interactive, more like a social network[45]. On the contrary, an artificial neural network is almost 100% learning from a previous layer. It does not have a functional feedback mechanism like the human nervous system.

The human brain makes inferences from this neuronal feedback mechanism, and relays that information directly to other parts of the system without using a "backpropagation" signal methodology for entrainment like an artificial neural network does. This inefficiency in artificial neural networks is one of the most important missing elements in artificial intelligence development.

From that standpoint, artificial intelligence is many years away from being able to mimic the human brain... That is, of course, unless there could be a quick workaround to this inefficiency in the feedback process that has stalled A.I. advancement.

What kind of solution can be presented to fix this problem? How can an artificial neural network be made more comparable and compatible to the human mind?

Our answer lies in the cybernetics approach to neuroscience that we discussed previously. Bateson's theory of mind referred to a "hierarchy of logical types", or a "hierarchy of contexts". These contexts are a type of meta-communication inherent to the neurotypical human mind, and are very much related to the feedback mechanism in human intelligence.

Outwardly, these meta-communications appear as context cues, such as allegory, symbolism, sarcasm, body language, and other subtle social cues that the neurotypical mind would pick up on and understand. However, an artificial neural network lacks this feedback mechanism, and therefore has a reduced capacity of this hierarchy of contexts. Likewise, people with certain psychological conditions, such as autism, also have a more shallow hierarchy of contexts than neurotypicals.

This reduced hierarchy of contexts in people with autism can be explained by a model of perception known as a Bayesian brain model[46]. Bayesian decision theory is a principled description of the processes that enable observers to derive the most probable interpretation of the environment. The unique perceptual experiences of autistic people lead to a tendency to perceive the world more accurately, rather than modulated by prior experience. In other words, they are less biased by previous experiences.

The Bayesian model of autism perception makes it comparable to A.I. learning, in that the lack of a feedback mechanism in A.I. mimics the Bayesian decision making process (as a result of the disregulated feedback loop common to many psychological disorders) in autism.

Therefore, using Bayesian decision theory to replicate autistic perceptual processes in A.I. can theoretically function as an effective workaround for the feedback problem inherent in neural networks. This can lead to more rapid advancement of A.I. applications.

IARPA is currently funding a project called MICRONS[47], whose goal is "to mesh neural networks with existing models of neuroscience research, and find that "secret" algorithm that makes the brain capable of "training" with far less input using a predictive model."

Designing algorithms that mimic autism learning through Bayesian decision making can allow A.I. to function in this capacity. The strong visual learning style associated with autism can be used to enhance A.I. Additionally, neural networks' lack of feedback learning makes it easier to control.

This model fills the bill perfectly for IARPA. This can make neural networks faster and more efficient. It is an established predictive model. An autism learning model can help A.I. to reach this short term goal.

In conclusion, it can be said that the similarities between A.I. and autism are being used as the perfect nexus point for the implementation of neuroprosthetics (computer-brain interface technologies), and the coming transhuman singularity. Brain mapping via nanotechnology[48] is being used to engineer non-invasive brain interfaces with machines.

It would appear that, in addition to the development of advanced artificial intelligence systems, there is also an ongoing agenda to re-engineer mankind to become more compatible for machine interface. My speculation is that this process is being observed and measured by the dramatic uptick in psychological disorders, such as autism, Alzheimer's, schizophrenia, depression, bipolar disorder, ADHD, etc. Approximately 1 in 5 people in the U.S. experience mental illness in a given year[49]. This number is staggering, and, I believe, indicative of experimentation on the public by a power structure that seeks total control.

So, now that we've shown how the future of artificial intelligence is intrinsically linked to autism, and we've outlined the basic concepts that demonstrate how the cybernetics approach to neuroscience can be used to induce autistic symptoms within the human nervous system, we should probably move beyond just discussing the concepts, and begin to establish the methods being used to inflict this agenda on the public.

This agenda that we're speaking of is the transhumanist agenda, and it is using the autism epidemic (and artificial intelligence) as a vehicle to force the next step in human evolution[50] through guided evolution.

Think this is starting to sound crazy? Stick around a few years. The transhumanist singularity is coming, but it will leave widespread destruction in its wake. Let's connect some more dots to better understand where this going, and how it is being pulled off.

Let's take a comprehensive look at the delivery methods being used to force the transhumanist technologies upon us, and how they are inflicting autism and other mental disorders upon us. These disorders are not just random symptoms that have always afflicted mankind at this scale. There is a massive eugenics campaign going on behind the scenes with this.

Then we will examine the social engineering strategies that they are using to normalize these conditions and bring about the demand for transhumanism as the answer to these problems that face us. The public is being conditioned to accept the gradual encroachment of ever more invasive surveillance technologies, as well as the acceptance of massive illness as the new norm for society.

Let's go ahead and examine a few of the key delivery methods involved in making the public sick, and attempting to re-engineer nature into the ultimate artificial panopticon control grid...

Chapter Twelve - Possible Causes of Autism and Methods of its Production

So, what have we been able to learn about what causes autism? There are as many different answers as there are approaches to the problem. There are many different facets to this epidemic, and seemingly endless little clues that lead professionals from many diverse research fields down numerous fruitless branching paths, never getting any closer to an answer. That is, until they take a step back and look at the whole picture. Examine it from the cybernetics whole systems approach.

Is there really something to this cybernetics approach to the autism epidemic? Or is this whole idea just this author's opinion? The answers to these two questions are, first, yes, there really is something to this approach, and secondly, no, the holistic cybernetics approach to the autism epidemic is not merely just my opinion.

Dr. Richard Tsien, chair of physiology and neuroscience at New York University, in a recent interview[51] cited colleagues who came up with a unified theory of autism's underpinnings in 2016. This theory verifies much of what was layed out in previous chapters of this volume, especially in regard to homeostatic disregulation and feedback loops. This theory[52] offers a holistic systems approach, in the vain of cybernetics, as a means of understanding the mechanisms that trigger autism.

It has been demonstrated in numerous studies that disregulation of neuronal calcium channels has been implicated in many different brain conditions[53], such as autism, schizophrenia, bipolar disorder, and depression, among others. This may be the key to understanding the causative mechanisms that trigger autism. If disruption of calcium channels in neurons has been implicated, then the question becomes, what causes the disruption of neuronal calcium channels?

There are several synergistic factors that can be contributing to this disregulation of neuronal calcium channels. And, the answers to these questions are not going to be comfortable ones, as the implications of these answers would seem to indicate that there is an interdisciplinary, purposeful, well thought out, deliberate, negligent and/or nihilistic agenda at play.

Let's begin by looking at what is on everyone's mind right about now, what could probably be the biggest factor commonly correlated to autism, vaccines. Are there any substances in vaccines that can cause disruption to neuronal calcium channels? Not all vaccines have the same ingredients, but almost every vaccine has at least one ingredient in it called an adjuvant.

An adjuvant is an ingredient in a vaccine that is there to agitate your immune system to enhance your body's immune response to the antigen in the vaccine. Let's take a look at these adjuvants, as this would be something common to all vaccines.

What is the most commonly used adjuvant in vaccines? Aluminum is the most widely used adjuvant in vaccines here in America. So, it would stand to reason that if aluminum can cause disruption in neuronal calcium channels, then, by proxy, vaccines would also have the potential to cause disruption in neuronal calcium channels.

So, does aluminum have the potential to cause disruption of neuronal calcium channels? Yes, there are numerous studies (we will cite 4, but there are many, many more) that show that aluminum toxicity causes a homeostatic disregulation of neuronal calcium channels and calcium mechanisms[54,55,56,57].

Vaccines are only one component of a synergistic convergence of factors that are contributing to the deliberate production of the autism epidemic. Two other important infection vectors we will look at are electromagnetic frequency radiation and geo-engineering programs (commonly called chemtrails).

Is there evidence that electromagnetic frequency radiation can affect calcium channel homeostasis or contribute to autism? Yes. Once again, there are many different studies (we will cite two) that show these correlations[58,59].

This isn't even taking into account the upcoming rollout of the new 5G network. Once 5G is implemented en masse, I predict we will see another exponential jump in neurological disorders, as well as cancers.

Lastly, we will look at "geo-engineering" programs. Within the last year, these programs have been admitted to in the mainstream after many years of denial of their existence. People have commonly called this chemtrails. There are indeed real spraying programs going on, this is not a crazy conspiracy theory. One of the chief proponents of these programs is named David Keith, from Harvard University. The stated purpose for these geo-engineering programs is to try to "reverse" the damage caused by runaway climate change.

So, what kinds of substances are they spraying in these programs[60]? The most common substances associated with chemtrail spraying are strontium, barium, and aluminum[61]. Once again, we can see another synergistic layer of what is going on here.

Our environment is absolutely inundated with aluminum nanoparticles. Aluminum toxicity is likely the key trigger causing the autism epidemic. It can be shown that the evidence is overwhelming that aluminum toxicity is a real problem.

Why would the powers that be seemingly ignore this aluminum toxicity problem? Why would they allow this synergy of causation to inflict autism on the scale that it has? Haven't they noticed this correlation that is being pointed out here? Are they really that incompetent? Or is there something else going on?

My speculation is that they are knowingly experimenting with different substances on the public. They are targeting these substances at different specific parts of the causal circuit of our nervous systems in an attempt to create "autism without intellectual impairment"[62] in order to filter out "undesirable" personality traits, while fostering savant type skills. They are trying to re-engineer humanity to be more compatible to be merged with machines, and the autism epidemic is their gateway to do so.

Autism without intellectual impairment, this is the goal, and there is a massive social engineering acceptance campaign going on to try to normalize autism. They are also rewriting history to make it appear as if autism was with us all along. The next step in our journey is to explore the social engineering agenda to normalize autism, and make it into a desirable condition. Then, later, we'll examine how this all relates to the creation of "super soldier" programs.

Does this sound ridiculous? Stick around awhile. We're only just beginning to expose all of the many diverse facets of the agenda that is engineering the autism epidemic...

Chapter Thirteen - From Awareness to Acceptance - A Normalization Campaign

The next aspect of the autism epidemic we are going to look at involves social engineering of the highest order. There is a "normalization" campaign underway to bring about the acceptance and celebration of autism as the natural progression from autism "awareness". In fact, as I am writing this book, a new development on this front has occurred. Apparently, April is now no longer "Autism Awareness Month", it is now "Autism Acceptance Month", complete with a slogan, "Acceptance Is An Action".

There is even a website[63] set up solely for the purpose of promoting Autism Acceptance Month. This, of course, functions as a vehicle for shifting public perception of autism from a position of pity and tolerance (that was associated with "autism awareness") to a position of respect and a sort of "reverence". This website and campaign was set up by a non-profit called the Autistic Self Advocacy Network.

This is a perfect example of the social engineering tactic known as the "Overton Window".

This concept[64] is used to shift the public consciousness to accept as normal more and more bizarre ideas gradually over time. By introducing a new socially unacceptable idea, they shift the perception of what people accept as normal.

As this pertains to the autism epidemic, it looks something like this. First, autism is given recognition as a separate disorder from schizophrenia, most of the people who suffer from this condition are institutionalized.

The window shifts and autism is recognized as a spectrum disorder, families begin to understand and deal with this condition at home, no institutionalization. Next, the window shifts again as the autism rate spikes, and it becomes a far more visible disorder, new social services are put in place to help with care and treatment. The window shifts again to the concept of the autism awareness campaign. Now, we are actively watching the Overton Window shift again to the policy of autism acceptance.

The public perception of autism is beginning to be shifted into the direction of not just accepting autism as normal, but also accepting autism as the next step in human evolution.

Does this sound ridiculous? This is not my idea, there is a well documented campaign pushing autism as the next step in human evolution. This is exactly how the Overton Window works, it introduces an outlandish idea like this to make autism acceptance look way more reasonable by comparison.

But, I assure you, the push towards autism as the model for the next step in human evolution is coming, in fact, that concept is already well underway within academic circles as the study of "rhetorical" models of autism. We will take a look at some of these rhetorical models of autism a little later. There are even peer reviewed scientific studies suggesting that autism has had a very important role in the evolution of modern man[65].

Clearly, any one who is paying attention and questioning these things can see that there is a definite agenda at play here. What is the ultimate goal of this agenda? Why would the social controllers want to switch the public perception of autism from that of a disability to that of an evolutionary advantage?

The ultimate goal they are planning for is the transhumanist singularity, the merging of man with machine[66], the ages old quest for immortality, the attainment of the "Philosopher's Stone", the achievement of the "Great Work".

The transhumanist agenda has its roots in the ancient mystery schools of antiquity, whose teachings have been carried forward to this day within the various secret societies and fraternal brotherhoods of today, such as the Freemasons, Rosicrucians, Order of the Eastern Star, The Hermetic Order of the Golden Dawn, etc. The top most levels of these groups are commonly collectively referred to as the "Illuminati" among conspiracy researchers.

The perception of autism as an evolutionary advantage plays right into the idea that the many negative symptoms associated with autism can be "corrected" by using machine interfaces[67] as therapies to allow only the manifestation of "autism without intellectual impairment". They will use the guise of the treatment of autistic symptoms with augmentative therapies as a means of pushing forward the idea of the transformation of the human into human plus, the transhuman, the next step in human evolution, and it will be self-guided evolution.

Cognitive augmentation and physical augmentation via transhumanist technologies will be the ultimate treatment plan in order to correct the problematic symptoms of autism. This is the means by which the transhumanist singularity will be achieved. Those at the top of the power pyramid are using those afflicted with autism in all of its many iterations as GUINEA PIGS to test out these new technologies that they hope that they will be able to use to enhance themselves and live forever.

That is my estimation of the truth of the matter, the despicable "elites" who run the world are planning on using the autism epidemic as a grand experiment to test the waters of their new technologies. I assure you, that even though the transhumanist agenda presents its case as a benefit to all mankind, the so-called elite have no place for the "profane" (that's you and me) in their new age. They have no intention of allowing the common people to be the beneficiaries of these technological advances.

Instead, we have been scheduled for extermination, "the great culling". I encourage people to look up these terms. There has been an ongoing, many faceted, longterm eugenics[68] depopulation plan in the works for many years.

These are all steps in a long term plan, and we are rapidly approaching the endgame phase of this plan. We have been able to adequately show that there is a definitive normalization plan for the acceptance of autism. We have also explored the question of why.

The next logical step we should take in our journey is questioning how this is being done, and what tools are being used to push forward this agenda of autism normalization. In order to do so, we will examine the "rhetorical" models of autism that we mentioned earlier. We will also show how entertainment is being used as a medium to subtly push this idea on the masses.

Let's dive in to the "rhetorical" study of autism in popular culture...

Bazinga! - the pop culture view of autism

Chapter Fourteen - Rhetorical Autism and the Social Construct

What is the rhetorical model of autism? And, how does this concept affect the public perception of the autism epidemic in western culture? What are the societal implications that this impetus could have on our future? How deeply entrenched in our entertainment industries is this concept?

Let's start by first defining what rhetoric is. Plato describes rhetoric as "the art of enchanting the human soul". Francis Bacon says that the purpose of rhetoric is "to apply reason to imagination for the better moving of will". It could be said that rhetoric is an instrument of language used to frame an argument in a way so that the audience can understand it[69].

In this case, the rhetorical model of autism is understood to be the socially accepted perception of a stereotypical description of autism. This rhetorical model is portrayed throughout all of our entertainment and media in the guise of "geek" culture, technologically savvy people with poor social interaction skills. This is the perception Hollywood is pushing as the archetypal model of autism. This does not necessarily reflect the reality, but this is model that comes immediately to mind for anyone immersed in popular culture.

This rhetorical autistic archetype subtly endorses concepts like postgenderism, posthumanism, technocratic collectivism, and, most importantly, transhumanism. Some scholars[70] would also suggest that the rhetorical model of autism represents a reconfiguration of the last vestiges of masculinity (particularly white male hegemony) into a new technocratic form of asexualism, or postgenderism.

Once again, autism gets correlated to notions of gender confusion. This representation again lines up with the "extreme male brain" theory presented earlier. It would seem that the overall agenda coincides with the postgenderism push that will lead directly into transhumanism.

Let's look at some examples of pop culture characters that subtly represent the autism community. These characters are generally portrayed with many of the same characteristics. Almost every single autistic character portrayed in media is male. Usually, they will be highly intelligent. They will be undeniably socially awkward. They often lack empathy or emotion. They very often show an affinity for science and technology. They often show little or no interest in forming relationships with the opposite sex. They tend to have a strong bond with another neurotypical male character. The stigma associated with autism is generally not applied to these characters, they are held in high regard for their skills.

Let's list some examples of characters in pop culture franchises that meet these criteria, and represent the rhetorical view of autism:
1) Mr. Spock - Star Trek (original series)
2) Sheldon Cooper - Big Bang Theory
3) Mr. Data - Star Trek TNG
4) Dr. Shawn Murphy - The Good Doctor
5) James Halliday - Ready Player One
6) Mister Fantastic - Fantastic Four
7) Raymond Babbitt - Rain Man
8) Dr. Isidore Latham - Chicago Med
9) Forrest Gump - Forrest Gump
10) Brick Heck - The Middle

That's just a partial list that gives a pretty good representation of what the rhetorical view of autism looks like. This mostly just applies to pop culture entertainment. However, there is a concurrent concept in the scientific field that is post-humously attributing an autism diagnosis to certain important historical figures. This is a blatant attempt to rewrite history for the purpose of making it appear as if autism has been with us throughout all of human history, and had played an important part in our evolution.

This whole idea screams of an agenda to me. Let's take a look at this push in the scientific community to once again normalize the idea of autism, and actually place it in a frame of reference of respect or reverence. We'll look at this idea that autism was an integral part of our development, and then we will show that the idea of autism has not been around from time immemorial, that it is only, in fact, a very recent phenomenon, dating back at its furthest traceable point to only 1911. Let's look at the historical figures that "experts" now claim were autistic...

Was Einstein autistic?

Chapter Fifteen - Rewriting History to Support the Narrative

The manipulation of information is a key part of the social engineering agenda. One way the social controllers do so is by seizing the avenues of information dissemination. Information distribution methods include news and media, textbooks, magazines and periodicals, technical journals, and entertainment, among other channels.

If you control all of these outlets, you can have a near monopoly on information. This is how the powers that be steer and direct the narrative that they want the masses to believe. There is power in the written and spoken word.

One example of how they set certain narratives, is the way they are actively rewriting history by posthumously attributing famous historical figures with autism or autistic traits, with little or no evidence to support these claims. Why would somebody be doing something like this when there is scant evidence to back up these claims?

The answer is simple. They are doing so because they have an agenda. Disseminating this information supports their agenda. What is this agenda? The normalization campaign of the autism epidemic.

Why would they want to normalize the autism epidemic? The answer to this question is another simple but profound one. They are doing this not only because they are expecting the autism rate to rise even more drastically than it already has, but also because it benefits them to have an epidemic level of autism in the world.

This sounds like a very sadistic thing to say, but, from their point of view, having larger numbers of autistic people around, on all different parts of the spectrum, gives them more statistical data to work with. Higher numbers of data points equate to higher levels of control. It's a simple matter of statistical analysis. You can produce more accurate models when you have higher quantities of data.

And, remember, their intention is to use those on the autism spectrum as proverbial guinea pigs to beta-test their neuroprosthetic devices and transhumanist augmentation technologies. It's really quite sinister.

So, what exactly are they hoping to accomplish by insinuating that autism has been with us all throughout human history, and has been an integral component of our evolution? They are trying to put a positive spin on autism. They are trying to create "autism without intellectual impairment".

They are not only trying to push autism acceptance, but they are also trying to turn autism into a desirable condition, or set of traits, and they firmly believe that they can do that. In fact, arguably the most famous living person on the autism spectrum, university professor, Temple Grandin, proudly states that she firmly believes that "characteristics of autism can be modified and controlled".

So, therefore, by posthumously accrediting autism to universally admired, important historical figures, they lend a certain sense of respectability to and desirability for the condition. Seeing this social engineering agenda for what it really is gives you a new sense of how we are all being used and manipulated by these purveyors of sociological narratives.

It is not easy to come to terms with the fact that the people who drive these agendas have no regard for you or me, or our autistic progeny. They are not driven by the same needs and desires as you and I. They are self-seeking narcissists who are driven by the overwhelming desire to control all aspects of the world around them, and live forever. Their intention is to become God through the advent of their transhumanist control grid, which they are actively setting up quietly all around us as we sit idly by and do nothing about it. Complacency is the plague that will be the undoing of our society unless we begin to wake up, speak up, and remove our consent from their system. And, keep in mind, they view your silence as your consent for them to control you.

Let's look at some examples of how they are attempting to rewrite history to prop up the narrative that autism has been with us all along, that autism was important to human evolution, and many famous important people were autistic. All of this is manufactured for the sake of making the public buy into this narrative so that no one tries to stop the autism epidemic, or the various technologies that coincide with it.

We can start by outlining some academic studies that claim that autism was an important feature in human evolution. There are several university studies positing that autistic people held very important positions in ancient pre-modern human societies, that they had special skill-sets that made them important to the survival of the tribe as a whole[71]. There is only some minor evidence to suggest a genetic link to autism may have been present back in that timeframe, but, as is the case with a lot of big moneyed interests, the agenda sets the narrative to line up with the social engineering goal of the controllers of that money.

This case is no different. You get a credentialed "expert" on the subject to basically invent the narrative out of whole cloth, and then throw in the backing of a major university, and ultimately publish these ideas in academic journals and call it "science". This is how a very few people control how the masses perceive reality.

In this case, the claim is made that autism was imperative to the survival of humanity sometime way back in prehistory, and that we should embrace autism because of this. The inference is also made that autism may be important to our further evolution into the post-human (or transhuman) era.

You can see how this "invisible government" steers and directs the illusion that is the public perception of reality[72]. These methods are often very subtle, and if viewed separately, they seem innocuous, but once you look at things from the holistic perspective, you begin to see patterns emerge. It then becomes obvious that someone somewhere is trying to convince you that something is factual, or real, and oftentimes without you even questioning their story because they are the "authority" in the subject.

The other aspect of this we can examine is the list of important historical figures who they are now saying had autism[73]. This whole idea is contrived to sway public opinion to accept autism as an important contribution to society. Let's take a look at a sizable partial list of some of the people that they now claim, posthumously, had autism.

Famous historical figures who allegedly had autism:
1) Hans Christian Anderson
2) Lewis Carroll
3) Charles Darwin
4) Paul Dirac
5) Albert Einstein
6) Thomas Jefferson
7) Alfred Kinsey
8) Stanley Kubrick
9) Steve Jobs
10) Nikola Tesla
11) Sir Isaac Newton
12) Andy Warhol
13) Wolfgang Amadeus Mozart

14) Michelangelo

This is just a partial list. As you can see, these are people who had a great deal of influence in their respective times and places. The inference that all of these people had autism diminishes the implications of the diagnosis. This is a grave disservice to those who struggle with this disorder every day. Yet, this is spun as a positive thing, rather than seeking a cure, they lobby for autism acceptance. That should give you an insight into the motives behind this social engineering agenda.

This agenda is designed secondarily as a means to transform our healthcare system from a system of disease management into an industry of beyond human cybernetic augmentation...

The Triangle of Enhancement Medicine, Disabled People, and the Concept of Health: A New Challenge for HTA, Health Research, and Health Policy

AHFMR

Policy implications

1. The number of "patients" increases dramatically.

2. The medicalization phenomenon accelerates.

3. A two-tiered healthcare and health system might develop: one dealing with the basics and one dealing with augmentative/enhancement medicine.

4. If one leaves the growing augmentative/enhancement field unregulated, without standards and supervision, one might see an increase in people becoming clients of the basic health care system due to botched procedures and side effects.

5. A brain drain toward the augmentative/enhancement cutting edge medicine might develop.

6. An ability divide will appear because many people, especially the traditional disabled people, will not be able to afford the enhancement treatment.

Chapter Sixteen - The Transhumanist/ Enhancement Model of Disability/Impairment

The institutions involved with the longterm planning of health and human services fields are anticipating and projecting even further escalation of the autism rate. Autism is a condition that legally classifies as a "disability".

Much like other disabilities, modern medicine is working on ways to provide new treatment possibilities through the use of technology in order to help patients attain better functionality. But, these new technologies require a planned infrastructure, complete with a way to pay for them.

Policy planning in order to accommodate these new miracles of modern medicine becomes an absolute necessity. New sciences bring about new ethical questions as well, so all this needs to be considered in policy planning for these technological advances.

Much of this planning has directly to do with prosthetics technologies. When you hear the word prosthetics, you probably think primarily of things like artificial arms, or legs. The truth of the matter is that prosthetics technologies in the near future will be way more than that. Let's define the term "prosthesis" to better understand what we're talking about.

"Prosthesis[74], artificial substitute for a missing part of the body. The artificial parts that are most commonly thought of as prostheses are those that replace lost arms and legs, but bone, artery, and heart valve replacements are common, and artificial eyes and teeth are also correctly termed prostheses.

The term is sometimes extended to cover such things as eyeglasses and hearing aids, which improve the functioning of a part. The medical specialty that deals with prostheses is called prosthetics."

So, you see, the term prosthetics could be used to describe the enhancement of a part of the body that does not function at a typical level. This has to do with much more than just replacing limbs, as we will see as we progress further along.

We need to take a careful look at another term that we all take for granted in order to see the direction that this is pointing to, and that term is "health". Health can be considered an umbrella term for a state of physical, mental, and social well-being. It can also be defined as the absence of disease or infirmity. So, this can potentially create a blurred line as to what it means to be "healthy".

If you consider the very flexible meaning of what the term "health" could represent, it tends to lose an important boundary of what should be considered "normal".

Likewise, when you consider the loose definition of what a prosthesis is, you can begin to see how some type of a medical enhancement could be argued to be for the maintenance of "health".

The impact of the ability of science and technology to improve upon and modify the human body (and mind) to "beyond species-typical boundaries" will lead to the inevitable demand for transhumanist augmentation and enhancement[75].

Eventually, our medical model will shift from one of disease management to an augmentative/ enhancement model.

Major healthcare organizations are actively planning for this paradigm shift in health policy. They are expecting the transhumanist movement to have a major impact on the medical industry.

This new approach to health and technology will bring about unprecedented changes in our society.

So, you're probably wondering, what does this have to do with autism? To get a clearer picture, let's look at a specific type of prosthetic device that shows a lot of promise in the future for treating all kinds of neurological disorders, such as autism.

This device is called a neuroprosthetic. Neuroprosthetics are machines intended to enhance brain function and control. The most talked about device of this nature is called "neural lace", and is activately being simultaneously developed by several tech companies, the most famous of which is Elon Musk's Neuralink.

Another company developing a version of this technology is called Kernel.

What exactly is "neural lace"? Neural lace[76] is an electromagnetic mesh that fuses with the brain to create an intermediary interface layer between the brain and a computer. This is a method to merge the minds of men with an artificial intelligence.

Can you imagine the implications of having your mind directly linked to the internet? This is the kind of thing we're talking about, the enhancement of human intelligence to the point of encompassing the collective knowledge and experience of all mankind.

The whole point is that these technologies will first be made available to those with disabilities, in order to restore typical functionality, and maybe even beyond typical functionality. This would inevitably lead to the demand for these transhuman augmentations to increase within the general public.

Once people begin to benefit from these beyond human capacity enhancements, it will create a large disparity in our society. Normal biologically typical and neurotypical people will be at a disadvantage to the transhumans, and therefore will be considered disabled, feeding this whole newly transformed healthcare system.

The healthcare industry is preparing for this inevitability. They are writing policy recommendations to deal with this transhumanist transformation of the medical field. It will come to the point that people will seek these modifications for aesthetic reasons, and not just out of medical necessity.

This is all spelled out very succinctly in a 2005 publication called, "The Triangle of Enhancement Medicine, Disabled People, and the Concept of Health: A New Challenge for HTA, Health Research, and Health Policy". This document was published by the Alberta Heritage Foundation For Medical Research's Health Technology Assessment Unit. Policy papers like this one set the stage for what is expected to come in the future.

The neuroprosthetic concept shows promise for use to treat autism symptoms[77]. In its beginning phase, it will probably be considered to be something akin to a pacemaker for the brain, but will very quickly develop beyond that simplistic design.

I speculate that in the very near future, we will see a company put out a neuroprosthetic device specifically designed for the treatment of autism. This will be the first step toward the transhuman singularity.

The major policy implications of this change in the medical model will play out in a progressive fashion. Whether you agree with the transhumanist enhancement/ augmentation model of health or not is irrelevant, the fact remains that this is a very likely probability for the future of medical technology.

As we stated here earlier, the intention that the controllers of these technological advances have, is to use those affected by autism as guinea pigs in this grand experiment. When it comes down to it, the people in charge of pushing the transhuman agenda do not have the best interests of the public in mind. They want to live forever and have absolute control of everything.

So, now we can see the very dangerous direction that this is all going. There is a really fine line between offering hope to people and blatantly using them to achieve a goal that doesn't align with their best interests. There would seem to be a lot of strange inexplicable connections between autism and this transhumanist movement.

Why is that? Why has the transhuman agenda latched on to autism as a means to an end? What is the reason why they are using the autism epidemic as a stepping stone in their plan?

As strange as it sounds, this all ties back to the military industrial complex and its ongoing quest to create a "super soldier". That's right, we're going to look at autism from the viewpoint of the military industrial complex, and its intelligence community assets.

Why are autistic people a desired commodity for military and intelligence agencies? Let's explore the "super soldier" link to the autism epidemic...

DARPA's T.A.L.O.S. Suit

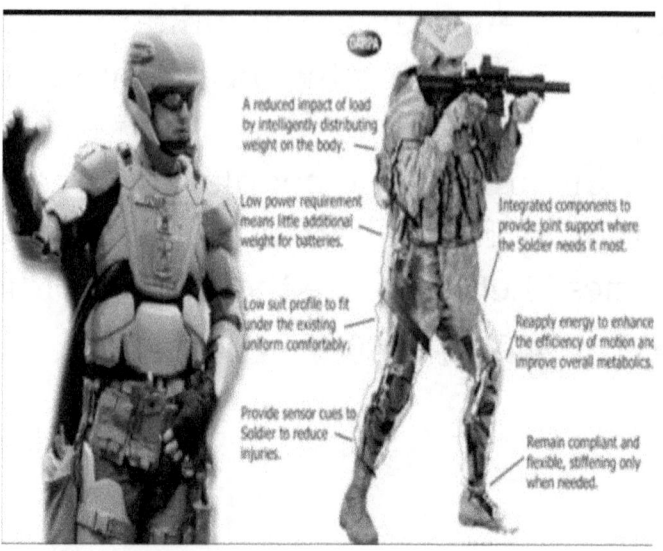

Chapter Seventeen - Autism and the Military Industrial Complex?

It is an almost inarguable fact that any new technology first finds its development through the auspices of the military industrial complex and it's related agencies. Indeed, the research and development of many advanced biotech products is funded by DARPA and other quasi-military organizations, or their associated corporate subcontractors (i.e. - Lockheed, General Electric, Raytheon, etc.).

Other projects are funneled through special quasi-governmental thinktank groups, such as the RAND Corporation or the Brookings Institute. Sometimes, some of these classified or sensitive projects are studied and developed by a special group of academics, scientists, engineers, and researchers from different disciplines called the JASON group.

It has been flatly stated from some of these project insiders that the military industrial complex's classified "black" programs deal with technologies and sciences that are at least 30-50 years ahead of what we would consider "state of the art" in the public sector. This is important to remember for two important reasons.

The first reason is this, if you see a new technology rolled out for public consumption, you can bet your bottom dollar that it has already been put through its paces first by the military industrial complex.

The second reason this is important for our discussion here today, is because the sciences and technologies we will be discussing have likely been around in these secret "black" projects programs for at least the past thirty years.

This means that these "future" technological developments most likely already exist, and have been developed and tested. Therefore, many of these "hypothetical" technologies are not just mere speculation, they exist now, and have existed for a long time in the classified world.

When you see these technologies discussed in public forums, rest assured that this means that the military industrial complex has already mastered the applications of these technologies for militaristic purposes.

Whenever a new development occurs, the military are quick to latch onto it and weaponize it. This is true of almost any scientific field of endeavor. It only gets released for public consumption after the military industrial complex has worked out every possibility for its use. Then, these technologies get released to the public.

I would speculate that there are certainly some technologies that the military industrial complex does not let come to the public sector. These would be technologies that would remain classified for security reasons, or are still being continued to be tested.

So, we can say that there probably exist sciences and technologies that we know nothing about, and probably can't even comprehend.

Let's step away from speculation for a moment and look at what types of technologies are being developed that relate to what we're talking about. One of the key scientific disciplines being studied in relation to creating "super soldiers", is neurobiology.

Neurobiological technologies are being developed in order to combat fatigue, stress, and distraction for soldiers[78] on the battlefield. One of these technologies applied in this way is called repetitive transcranial magnetic stimulation.

This technology has been shown to effectively enhance certain skills in people while it is employed. In fact, it has been demonstrated that this method can be used to temporarily induce "autistic savant" skills[79] in neurotypical people. This is a form of non-invasive brain stimulation.

Taking the next step in that vain, brain implants[80] are an integral part of developing a viable super soldier. Neural implants can be used as a computer-brain interface to control and operate military technology.

They can also be used to treat conditions like PTSD, or other neurological disorders. There is also the possibility of cognitive enhancement, such as the impartation of savant skills.This is a technology that is being pursued by the military industrial complex.

With the advent of these technological cognitive enhancements, the military needs just the right kind of people to augment with these technologies to achieve maximum performance. This is where autism comes into the picture, more specifically, the higher functioning forms of autism (like Asperger's).

High functioning autists have a lot of desirable traits for a "super soldier". Let's look at a list of these different factors. This once again relates to the goal of "autism without intellectual impairment".

16 autistic characteristics that are desirable for a super soldier:

1) Enhanced senses - many autists have hypersensitivities

2) Follow orders without concern for others- many autists lack empathy

3) Enhanced spatial awareness - autistic intelligence is highly visual (VPR model of intelligence)

4) Reduced need for sleep - most autists' bodies do not naturally produce enough melatonin, causing them to sleep much less than average

5) Specialized skills - high rate of savantism among autists

6) Photographic memory - many autists have a very good memory and retain visual information better

7) Attention to detail - many autists have high visual acuity

8) Ability to use technology effectively - many autists have an affinity for technology

9) The ability to perform duties without distraction - many autists will hyper focus on things of interest

10) A "metabolically dominant" soldier - many autists have a fast metabolism and heal quickly

11) High intelligence - autism is described as a disorder of high intelligence

12) Enhanced strength - hormonal stressors can trigger short bursts of increased strength in those with autism (often called an autistic "meltdown") - a means to control these short bursts of strength could be invaluable to the military

13) Increased compatibility with A.I. - autism intelligence is a model for artificial intelligence - the ability to merge an A.I. machine with the autistic brain through a computer-brain interface is the future of warfare

14) A harmless appearance - someone who doesn't look like a threat is a highly desirable operative for military intelligence

15) Ability to compartmentalize tasks - autistic people are known for compartmentalizing and categorizing things

16) Ability to follow a structured chain of command - autistic people have a need for routine or structure.

We can see that autistic traits can be ideal for developing a super soldier. But, at a militaries actually seeking out autistic recruits? As strange as it sounds, yes, they are. Many military and intelligence organizations actively recruit autistic individuals to work for them[81]. The same is true for computer tech support and IT companies[82].

Many autistic individuals excel in these types of jobs, as well as engineering jobs. Military and intelligence agencies often look for autistic people for their computer skills. They are often recruited as hackers and drone operators.

Future military operations will be less about boots on the ground, and more about effectively utilizing advanced technology, and that is where the autism link ties in. They are conditioning autistic people to be dependent upon technology to function in society.

This is the methodology that they use for treatment options for autism. In turn, autistic people become experts with these technological interfaces, and therefore make for desirable employees for tech companies and assets for military and/or intelligence organizations.

All of these different factors would tend to suggest that there is a concerted effort going on to promulgate the autism epidemic rather than find a cure or preventative measure for it. We have demonstrated that, through the holistic cybernetics approach, it has long been known what generally causes autistic symptoms, the disregulation of neuronal calcium channels in the nervous system. But, despite understanding how this general homeostatic mechanism of the body works, the question remains, can we point to just one or two factors that contribute to, or cause the autism epidemic?

Likewise, we've now examined the agenda claiming autism has been with mankind from the beginning of civilization, and also the agenda claiming autism is the next step in human evolution. But, can we find a specific point in time where autism emerged from?

Can we demonstrate that autism is only a more recent phenomenon?

Stated plainly, can we pinpoint when and how the autism epidemic started, and what causes it? The more difficult question is, can we show what the intent is behind this whole thing?

We will examine these questions and more. The next aspect of this epidemic that we'll look at is the emergence of the word autism, and the etymology of the word. We will examine it's very first recorded use, and the earliest known research into understanding the condition, as well as its probable causes, both then, and now.

Afterward, we will show some evidence of a possible covert project that may have something to do with the rise of neurological illness in our society.

Let's jump right in and analyze the origins of the autism epidemic...

Dr. Leo Kanner

Chapter Eighteen - The Origins of Autism

The very first use of the word autism came in 1911, when Swiss psychiatrist, Eugen Bleuler, used it to describe a specific set of symptoms in schizophrenic patients. It was originally stated as "infantile autism" because it described recessive behaviors.

The etymology of the word autism[83] comes from the Greek "autos" and "ism", meaning "isolated self", or "separated one". It was used to describe the odd behavior of patients who seemed to be in their own little world. They displayed a marked lack of normal social interaction skills, and thus this term was used to describe this subset of behaviors.

In the late 1930s and early 1940s, Dr. Leo Kanner (of John's Hopkins University fame) began studying children with unusual emotional and social behaviors. He later adopted the term autism to describe the clinical condition. Simultaneously, in Germany, Dr. Hans Asperger identified a similar condition that would be named Asperger's Syndrome.

Despite autism being identified as a unique condition, most researchers still categorized it as a form of schizophrenia right up into the 1960s. Medical professionals really didn't have an understanding of autism as a separate condition from schizophrenia until that time.

The era of the 1960s and 1970s were a rather horrific time for autism treatments. They experimented with things like electroshock therapy and LSD to try to treat autistic symptoms. The vast majority of moderate to severe autism patients were institutionalized

In the 1980s and 1990s, the medical community started to gain a little bit better understanding of autism. At this point, behavioral therapies and controlled learning environments began to become the norm for treatment. By the mid 1990s, the autism rate began to skyrocket, and continues to skyrocket right up to today.

So, now we know that the earliest historical record dates back to 1911. This condition did not exist up until that time. It is a logical fallacy to say autism has been with humanity throughout all time. Can we somehow correlate this time in our history (1911) to some type of a major change in society, or our environment that could possibly coincide with the emergence of autism?

If so, could we show that this change in our society and/or environment could possibly have something to do with the emergence of autism? We will begin to explore these questions in earnest to demonstrate that there is indeed a plausible answer that satisfies the criteria we are looking for, a more than coincidental link to autism.

In order to find an answer to this dilemma, I first had to ask the question, what do we know about this era of history? The late 1800s and early 1900s were the beginning point of the industrial revolution. The next logical question from there is, was there some type of industrial enterprise that sprung up around this time that could possibly have polluted the environment with something that could potentially cause these symptoms in people?

After I began to explore the ins and outs of these questions, I came to surprisingly simple yet profound answer to this inquiry. A cursory glance at the industrial advances of that time pointed to a clear and obvious answer to this dilemma. Let's look at a historical overview of a specific industry that arose around this timeframe, and we should be able to better understand how we got to where we are today.

In 1808, an English chemist named Humphry Davy discovered that, in theory, aluminum could be produced by electrolytic reduction from alumina (aluminum oxide). However, this was not able to be accomplished until 1825, when Hans Christian Oersted of Denmark, was able to produce an aluminum alloy in experiments.

Aluminum production[84] progressed very slowly, first through a German chemist named Friedrich Woehler, who after nearly twenty years of continuous experimentation, was finally able to produce a small amount of aluminum globules in 1845. This was eventually dubbed the "Woehler Process".

Throughout the next 45 years, aluminum production was too expensive and inefficient for any practical purposes. Improvements in the Woehler Process were made, but large scale aluminum production was still not a feasible thing.

The development of aluminum changed in 1886 with the invention of a new, much more cost-effective electrolytic production method. This method was called the "Hall-Heroult Process", named after the two individuals who simultaneously discovered this method. Charles Hall was an American student, and Paul Heroult was a French engineer. The process they invented produced great results, but required a vast amount of electrical power.

Even with these advances, aluminum production was still very slow and inefficient until 1889, when Karl Josef Bayer, an Austrian chemist invented a cheap alumina (aluminum oxide) production method. At this point, alumina became the basic raw material for aluminum production. Aluminum production today is based upon the Bayer and Hall-Heroult processes.

The aluminum industry began to evolve slowly over the next several decades. It began to take off shortly after the turn of the century in the early 1900s. After this, the aluminum production industry began to accelerate and grow rapidly, as aluminum was a key material for use in "wartime" products.

So, we can see that the rise of the aluminum industry coincides with the emergence of autism in the historical timeline. But, is there a reason for this? Does aluminum production produce harmful waste products or pollutants that can cause possible neurological side effects?

Yes, atmospheric pollutants from primary aluminum production can cause "acid rain", and contaminate water and soil with aluminum[85]. Aluminum in high doses is toxic to humans, and has been indicated in correlation to autism and Alzheimer's. The industry claims that as long as th pH level of soil remains at or above 5.0, aluminum concentrations should be harmless.

So, that brings us back full circle from where we started. Aluminum has been implicated as a probable cause for both autism and Alzheimer's. If a simpleton like me can figure this out, then doesn't it stand to reason that professional scientists and academics should be able to come to these conclusions also? Why do they allow this to continue and do nothing to stop it?

The answer to this question is profoundly simple, and disturbing. It is because the people at the top of the power pyramid who control this "science" not only know this is going on, but they also willfully allow it to continue for the fulfillment of an agenda.

This agenda is twofold. It is primarily the transhumanist agenda, and secondarily, a DEPOPULATION agenda. The elitist power circles after World War 2 began to become concerned with runaway population growth. After the "baby boom" of the late 1940s and 1950s, they began to fear that population growth would lead to depletion of resources, and a population reduction plan would have to be executed.

It was decided that a militaristic approach would be far too costly and messy, so they opted instead for "soft kill" methods, and social engineering tactics. The "great culling" of the masses is currently underway...

DTFN Estimates for Nano Domestic Quell
Phase 4 Updated Compliance

June 2013

Assistant Director of Advanced Projects
The Office of DARPA Command
For DEPSECDEF EYES ONLY

68-123

Nano Domestic Quell
Revised Estimates for NDQ Protocols

National Nano Domestic Quell (NDQ) Protocols for Phase 4

DTFN Estimated Rates & Phase 4 Updated Compliance for N.D.Q.

Current total infection rate for United States general pop.: 87.2%

Projected infection for general U.S. populace by January, 2014 is estimated to reach 98%. Total infection for ages 18 and above may reach 99%. DTFN projects dispersal mediums will require additional resources for Phase 4 of NDQ. DTFN recommends an increase in the following medium inflows and outflows, specific to liquid dispersal:

Pepsi Co: 9.9%
Nestle ADR: 8.5%
Chicago Municipal: 5.1%
Atlanta Municipal: 4.4%
Danone: 4.2%
Coca-Cola: 4.1%
Los Angeles Municipal: 2.9%
Seattle Municipal: 1.0%

Dispersal outflows have shown significant improvement in population infection rates.

Recommended inflow increases deployed in October, 2012 resulted in a net increase of infection rates by 0.82%, slightly exceeding projections.

DTFN assures DoD compliance for Phase 4 will be completed one week ahead of schedule.

No further recommendations have been submitted by DTFN for Phase 5. An expected update to outflow estimated rates will be forthcoming before Phase 5 initialization.

Approved for Release   Pg. 1

Chapter Nineteen - Nanotechnology, the Means to an End

The population reduction campaign has been ramping up the past few years. One of the key tools being used to implement population policies is nanotechnology. This may seem like a rather stunning statement, but we are unknowingly inundated with massive amounts of nanotech products every single day.

Nanoparticulates are used in practically everything these days. Food, beverages, medicines, VACCINES, even in the air we breathe. In order to better understand what we're talking about, we should probably go ahead and define a few key terms.

What exactly is nanotechnology? Nanotechnology is science, engineering, and technology conducted at the nanoscale[86], which is about 1 to 100 nanometers. A nanometer is one billionth of a meter. For sake of scale, an average sheet of paper is about 100, 000 nanometers thick. This is the scale of extremely small things.

Nanoscience and nanotechnology involve the capability to control and manipulate individual atoms and molecules. This is engineering at an extremely small level, which has many advantages when compared to engineering something at the macroscale.

The engineering of nanomaterials has many practical applications. For the sake of what we're talking about here today, we are just going to concentrate on the applications that are relevant to biological systems, and the substances being used to manipulate the human body (and mind) at the nanoscale.

We are at the precipice of an era where nanoscience is making exponential leaps forward in its capabilities. Nanotechnological advances in biological and medical applications are an astounding example of just what can be done to biology on the macrolevel, by manipulating the micro- or nano- level.

We now have tools, like CRISPR[87], that can alter DNA sequences with relative ease, and those affected may not even know that this tool has been used on them. One particular use of CRISPR with the potential to be a grave threat is a technology called a "gene drive". A gene drive[88] can be used to alter the DNA of a specific target, and pass on the new genetic alteration to the original target's offspring. This technology has the potential to wipe entire species (or, familial bloodlines). There are currently no laws in place governing the use of gene drives.

Gene drives have already been used to modify insect species (specifically in mosquitos) in order to make them infertile. This technology works, and could be delivered to the target via many possible methods.

Gene drives and CRISPR are not the only concerning nanotechnologies that we need to keep an eye on. Even manufactured nanoparticles can have some undesirable effects.

One specific type of nanoparticulate that is now being found to contribute to various health problems is titanium dioxide (TiO2). It is used in many food products. Titanium dioxide[89] nanoparticulates are commonly in candy and sweets, dairy products, cosmetics, and a wide variety of other commercial products. Nanoparticulate titanium dioxide can be toxic to the human nervous system.

The International Agency for Research on Cancer (IARC) classifies titanium dioxide as a Group 2B Carcinogen, meaning it is known to cause cancer. Titanium dioxide is also commonly found in the very air we breathe.

Although the abundance of titanium dioxide in various facets of our environment is troubling, it is not the primary substance of concern in our review of nanomaterials. One of the most common elements used in nanoscale applications is aluminum. We've discussed some concerns with aluminum previously, but in nanoparticulate form, it becomes extremely troubling.

Aluminum hydroxide[90] is the most widely used vaccine adjuvant in the world. Recent advances have led to the production of this adjuvant at the nanoscale. The most troubling part of this is that it has been repeatedly demonstrated that metallic particulates at the size of 200 nanometers or smaller can freely cross the blood brain barrier and cause neurological or cerebral damage.

Yet, the pharmaceutical industry continues to develop and use various nanoparticulate forms of aluminum adjuvants[91] in its vaccines, ostensibly because these adjuvants "improve" immune response to the antigen.

And, although they admit more study needs to be done on the safety of these aluminum nanoparticles, they continue to use them, knowing full well that the potential exists for adverse reactions, and/or neurological or cerebral damage.

Vaccines are the primary contributor to aluminum toxicity in people. Other secondary factors are things like gross air pollution caused by chemtrail spraying (also known as geoengineering, solar radiation management, stratospheric aerosol injection, etc.), contamination of our food and water supplies, and many consumer products, such as deodorant and cosmetics.

The use of nanotechnology does not stop at just nanoparticulate matter used in consumer products[92]. There are very real technologies being tested upon the public unknowingly. Evidence of these nanotech test programs have been leaked to the public[93] through document dumps and other methods. We are talking about actual nanomachines, and not just particles.

Some examples of these technologies are openly discussed by military and intelligence agencies as well as their corporate subcontractors[94]. Examples of these technologies include the following terms:

Smart Dust
Neural Dust
Nano-tags
Nano-sensors
Dynabeads
DARPA's ElectRx (healing nanobots)
Micro dust weaponry

And there are many, many more examples of nanotechnological innovations that are affecting our biology. Many of these technologies are outlined in DARPA and NASA documents. Some of these nanoscale devices and programs are very concerning.

But, perhaps the most shocking and disturbing revelation of all comes from a classified portion of a leaked DARPA document back in 2013. Only two pages of this document made it's way to the public view, but the implications of this secret, experimental nanotechnology program implemented upon the American public without their knowledge or consent is absolutely horrifying. These pages indicate that 87% of the population had already been "infected" at this time, with an estimate that 98% of the population would be successfully infected by the end of January 2014.

This controversial document was titled "Nanodomestic Quell". The Department of Defense and DARPA to this day will claim that this document is a hoax, but based upon some of their tech programs that have been disclosed to the public, it is not outside of the realm of possibility that this is a legitimate document. After all, government agencies have been known to experiment on the public in the past[95]. This is an indisputable historical fact. What would stop such an agency from doing so again?

So, exactly what kinds of technologies would they be infecting us with[96]? And what would be the purpose? How come we wouldn't know that we had this technology in our bodies? Wouldn't there be some sort of signs or symptoms?

If these things exist, why would they seem to be inert? Could these be some sort of trojan horse bioweapon? Is there some sort of catalyst that could potentially trigger these nanobots to activate and perform whatever mission they've been programmed for?

Could we really be infected with nanotechnology just awaiting a specific signal to carry out its pathogenic mission? Is there evidence that we do have this technology in our bodies? We need to look no further the curious affliction known as "Morgellons Disease". This could be possible evidence of the presence of this technology embedded within our biology.

Alleged Morgellons patent

(12) **United States Patent**
Hogness et al.

(10) Patent No.: **US 6,245,531 B1**
(45) Date of Patent: *Jun. 12, 2001

(54) POLYNUCLEOTIDE ENCODING INSECT ECDYSONE RECEPTOR

(75) Inventors: **David S. Hogness**, Stanford; **Michael R. Koelle**, Menlo Park; **William A. Seagraves**, San Diego, all of CA (US)

(73) Assignee: **Board of Trustees of Leland Stanford University**, Palo Alto, CA (US)

(*) Notice: Subject to any disclaimer, the term of this patent is extended or adjusted under 35 U.S.C. 154(b) by 0 days.

This patent is subject to a terminal disclaimer.

(21) Appl. No.: **08/465,593**

(22) Filed: **Jun. 5, 1995**

Related U.S. Application Data

(63) Continuation of application No. 07/954,037, filed on Sep. 30, 1992, now Pat. No. 5,514,578, which is a continuation of application No. 07/482,749, filed on Feb. 26, 1990, now abandoned.

(51) Int. Cl.⁷ C12N 15/12; C12N 15/62; C12N 15/10, C12N 15/63

(52) U.S. Cl. 435/69.7; 435/69.1; 435/252.3; 435/419; 435/320.1; 435/325; 435/348

(58) Field of Search 536/23.5, 435/69.1, 435/240.2, 325, 530/350

(56) **References Cited**

U.S. PATENT DOCUMENTS

4,704,362 11/1987 Itakura et al. 435/253
4,818,684 4/1989 Edelman et al. 435/7
5,514,578 * 5/1996 Hogness et al. 435/240.2

FOREIGN PATENT DOCUMENTS

WO 90/06364 6/1990 (WO).

OTHER PUBLICATIONS

Ashburner et al. (1974) Cold Spring Harbor Symp. Quant. Biol., 38:655–662. The Temporal Control of Puffing Activity in Polytene Chromosomes.
Evans (1988) Science 240:889–895 The Steroid and Thyroid Hormone Receptor Superfamily.
Green and Chambon (1988) Trends in Genetics 4:309–314 Nuclear Receptors Enhance Our Understanding of Transcription Regulation.
Segraves (1988) Ph.D. Thesis, Stanford University Molecular and Genetic Analysis of the E75 Ecdysone-Responsive Gene of Drosophila melanogaster.
Knust et al. (1986) EMBO J. 5:891–897 The chicken oestrogen receptor sequence: homology with v-erbA and the human oestrogen and glucocorticoid receptors.
M. Kanehisa (1984) Nucleic Acids Res. 12:203–213 Use of statistical criteria for screening potential homologies in nucleic acid sequences.

Hershko and Ciechanover (1982) Ann. Rev. Bioch. 51:335–364 Mechanisms of Intracellular Protein Breakdown.
Miller et al. (1985) EMBO J. 4:1609–1614 Receptor zinc-binding domains in the protein transcription factor IIIA from Xenopus oocytes.
Freedman et al. (1988) Nature 334:543–546 The function and structure of the metal coordination sites within the glucocorticoid receptor DNA binding domain.
Severne et al. (1988) EMBO J. 9:2503–2508 Metal binding "finger" structures in the glucocorticoid receptor defined by site-directed mutagenesis.
Giguere et al. (1986) Cell 46:645–652 Functional Domains of the Human Glucocorticoid Receptor.
Danielsen et al. (1987) Mol. Endocrinol. 1:816–822 Domains of the Glucocorticoid Receptor Involved in Specific and Nonspecific Deoxyribonucleic Acid Binding, Hormone Activation, and Transcriptional Enhancement.
Rusconi et al. (1987) EMBO J. 6:1309–1315 Functional dissection of the hormone and DNA binding activities of the glucocorticoid receptor.
Mader et al. (1989) Nature 338:271–274 Three amino acids of the oestrogen receptor are essential to its ability to distinguish an oestrogen from a glucocorticoid-responsive element.
Umesono and Evans (1989) Cell 57:1139–46 Determinants of Target Gene Specificity for Steroid/Thyroid Hormone Receptors.
Umesono et al. (1988) Nature 336:262–265 Retinoic acid and thyroid hormone induce gene expression through a common responsive element.
Kumar and Chambon (1988) Cell 55:145–156 The Estrogen Receptor Binds Tightly to Its Responsive Element as a Ligand-Induced Homodimer.
Guiochon et al. (1989) Cell 57:1147–1154 Mechanisms of Nuclear Localization of the Progesterone Receptor: Evidence for Interaction between Monomers.
Picard and Yamamoto (1987) EMBO J. 6:3333–3340 Two signals mediate hormone-dependent nuclear localization of the glucocorticoid receptor.
Pratt et al. (1988) J. Biol. Chem. 263:267–273 A Region in the Steroid Binding Domain Determines Formation of the Non–DNA-binding, 9 S Glucocorticoid Receptor Complex.
Nauber et al. (1988) Nature 336:489–492 Abdominal segmentation of the Drosophila embryo requires a hormone receptor-like protein encoded by the gap gene knirps.
Oro et al. (1988) Nature 336:493–496 The Drosophila gene knirps-related is a member of the steroid-receptor gene superfamily.

(List continued on next page.)

Primary Examiner—Michael Pak
(74) Attorney, Agent, or Firm—Janet E. Reed; Saul Ewing Remick & Saul, LLP

(57) **ABSTRACT**

Polynucleotide sequences which encode ecdysone receptors have been isolated and expressed in host cells.

45 Claims, 3 Drawing Sheets

All of this may sound ridiculous, but these technologies do exist, there are patents on file outlining these devices. But, what does all this have to do with the autism epidemic? What is the catalyst for the activation of these nanomachines[97] if they are indeed infecting us? Are they able to communicate in some way?

This brings us to the next subject related to this topic, and its relationship to the autism epidemic. We will now discuss the implications of electromagnetic frequencies and microwave radiation, and the coming 5G network and its implications...

Alleged nanocommunications array

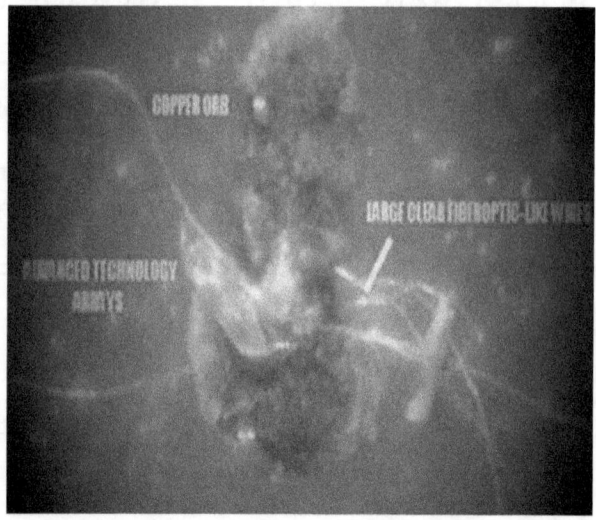

Chapter Twenty - EMF Activation of Nanodevices and the Singularity

Although this subject matter seems absurd, it is an undeniable fact that the technologies that we are outlining here absolutely do exist, here and now, documented in peer-reviewed scientific journals detailing these things. Many of these machines are openly talked about by futurists, like Ray Kurzweil, who disclosed that the invention of a nanoelectronic finite-state machine[98], or nanoFSM for short, would make it possible for the human neocortex to be connected directly to the Cloud.

This device[99] was invented in 2014, and is smaller than a human nerve cell. This machine has the potential to solve the conundrum of Moore's Law, the factor that limits the growth rate of computational power.

The ability to build entire electronic systems from bottom up within a biological form is not the stuff of science fiction. These nanomachines and nanocomputers can be programmed together in sequence to initiate various functions and effects when triggered by the proper external stimulus.

The external stimuli used to activate and guide these nanomachines are radio waves (microwaves), or specifically tuned pulsed electromagnetic frequencies. The infrastructure for this signaling apparatus is going into place as we speak.

The infrastructure we're talking about is the new 5G wireless network that is being rolled out in various large cities across America right now. This system, or its next generation counterpart, will be the catalyst to initiate the transhumanist singularity. This singularity is projected by experts[100] to happen sometime between 2029 and 2045. This is not a science fiction scenario, there are very real policy whitepapers that openly discuss the ramifications of this singularity.

When the singularity happens, we will all have our brains wirelessly connected to the Internet of Things via the Cloud. Within the Internet of Things will exist a subsystem[101] called the "Internet of Thoughts", linked through an artificial intelligence control grid.

And when this happens, mankind will have lost more freedom than can ever be imagined. Your mind will no longer be your own, you will no longer have free will. Your very thoughts (and pre-thoughts) will be monitored and shared with the "collective". At this point, man will arguably lose his soul.

This is the transhuman future, a collective hivemind containing all of humanity's thoughts, knowledge, memories, and emotions in a machine for all of eternity. There will be no more "self". No more "isolated self", the very definition of autism.

The bulk of humanity at this point in the near future will likely have been engineered to be mostly high functioning autists. The only exceptions will be those elite family bloodlines at the top of the power structure, who will be allowed to be "unplugged from the matrix". They will retain their individuality and free will.

These will be the controllers of this "matrix", the new gods of this artificial world of virtual reality and augmented reality.

This is what the transhumanist future looks like. Is this what we want for our children? To be enslaved to an artificial intelligence control system for all of eternity with no hope of escape or transcendence?

All of these technologies are being sold to the public as the new wonders of modern medical science. The brain-computer interface will be sold to the public as the solution to the autism epidemic. If you think that this sounds crazy, stick around for a few years, the transhuman singularity will be touted to be the "cure" of all of humanity's ailments, including aging. Doubt it not.

The EMF radiation emitted by the new 5G wireless network system will further exacerbate the health problems[102] that are already escalating at an alarming rate. The millimeter waves implemented by the 5G system are known and documented to cause irreparable harm to biological tissue.

The frequencies utilized in 5G are the same as those used by military "active denial systems". This is a weapon system[103] designed to burn the surface of skin to make it impossible for the target(s) to remain within the affected area.

If these frequencies are known to be harmful, then why use them? The answer is twofold. First, it all has to do with data input/output speed. They are looking to achieve faster speeds in order to allow communication at the speed of thought. The second reason is because they want to actively exacerbate the deterioration of people's health for the purposes of, first of all, reducing the population, and second of all, to use the classical social engineering tool known as "Hegelian Dialectic" to introduce the transhumanist singularity as the solution to all of the world's health problems. Hegelian Dialectic is a control system in which, you introduce a problem to incite a specific reaction from the public, and then present a preplanned solution to that problem. More simply put, problem, reaction, solution.

In this particular scenario, the problem is the accelerating health crisis, the reaction is the public outcry for a cure for these health problems, and the solution is the rollout of the transhumanist singularity.

To be more specific, the problem will be the unbelievable prevalence of the autism epidemic, the reaction will be a vociferous demand for answers and a cure, and the solution will be the introduction of neural implants to treat and reverse autistic symptoms. This will be the first step toward the gateway into transhumanism.

The autism epidemic is the most likely candidate to be the test run for these brain-computer interfaces that will herald in the post-human age. For some reason, it would seem that the elitist scientists at the top most tiers of the power pyramid like to use people on the autism spectrum as guinea pigs for their new scientific innovations.

The compatibility of the autistic brain with an artificial intelligence is probably the primary driving factor behind this line of reasoning. It would appear that those on the spectrum could be the first logical choice to be enhanced to become cyborgs. I know that sounds very much like fantastic science fiction nonsense, but there are numerous feasibility studies for the advent of cyborgs that have been done[104]. These are well documented.

In order to achieve this goal, the first requirement to do so is the massive collection of data. The next facet of this plan that we will examine is the new blockchain based data collection model being used by the NDAR, the National Database for Autism Research. This system serves as a model for a larger worldwide data collection program designed to quantify every piece of medical data of every person in the entire world...

Chapter Twenty-One - Data Collection and the Blockchain Trap

One of the most important actions that is necessary to advance the transhumanist plan is the collection of data, and more specifically, biometric and health data. The infrastructure for this endeavor is actively being built and developed. The goal is an international public ledger system for medical research.

This database will contain every piece of information about every individual alive now, and every individual who ever lived. It will be tied to a blockchain, so no information loss will ever take place. This data will be encrypted to protect an individual's identity.

Each individual will be identified by something called a G.U.I.D., a Global Unique Identifier. This will be a number that will identify you without revealing your name or personal information, but all of your medical and biometric data will be intrinsically linked to this number.

Ostensibly, anyone using this research database will not know your identity, but if you think that the artificial intelligence system attached to this public ledger won't be able to figure out exactly who you are, you're fooling yourself.

This international data repository will be modeled after the United States National Institutes of Health NDAR program. NDAR stands for National Database for Autism Research[105]. This system is built on blockchain, and has an enormous amount of data (approximately 500 terabytes) already collected on over 77,000 people on the autism spectrum in America.

Any time an autistic patient sees the doctor, this data is collected by the repository. Any social services that an autistic client receives, this information also finds its way into the data repository. Anytime an autistic person gets a prescription filled, this data is also collected by the NDAR. No not of data concerning the autistic patient goes uncollected.

Even school records, such as IEP's (Individual Education Plans), wind up in this data repository, attached to the individual's unique identifier number. This database also collects biometric information from patients. Health insurance billing and diagnostic codes are also captured.

This system is primarily used by medical researchers performing studies related to autism. They use this data to try to figure out all varieties of treatment methods, causal factors, behavioral therapies, etc.

They also focus heavily on genetic factors, and epigenetic factors. Several specific genes have been pinpointed as having a correlation to autism risk, as we discussed earlier. This database is one of the primary resources for those studies. This database is only for one specific condition.

The coming international database will cover every possible condition, illness, disorder, or anything else health related. Biometric data is one of the most sought commodities for such a system.

This collaborative database[106] will be managed by an organization called the Global Alliance for Genomics and Health (GA4GH). This group is a policy-framing and technical standards-setting organization, with a focus on the "responsible" sharing of genomics data within a human rights framework.

This collective has over 500 member organizations already onboard sharing data, including the NDAR. The infrastructure is in place for the mass collection of biometric and genomic data. The Global Alliance for Genomics and Health is an international non-profit organization formed in 2013 to create the framework for the collection of healthcare data.

An important technology that makes mass data collection a feasible accomplishment is the innovation known as "blockchain". Most people are familiar with the concept of blockchain because of the advent of cryptocurrency, such as bitcoin.

Blockchain can be used in much more innovative ways than for just cryptocurrency. Let's define what blockchain is so we can better understand how it will be used for biometric and genomics purposes.

Blockchain[107] is a digital, public ledger that records online transactions. It en add ures the integrity of data by encrypting, validating, and permanently recording data. Permanency is the key here, each new block in the chain carries over all previously recorded information along with the new information into the system, thus information can never be lost, changed, or erased. It creates a permanent record of all transactions on the chain.

This software application can be applied to biometric, genomic, and health data in order to paint a holistic overall picture of all aspects of the health and wellness of an individual. This can be used to create individualized treatment plans, and new avenues of medical research in order to understand the best options for individuals.

With all of the good that could be accomplished with such a utility, the potential for its misuse is of grave concern, especially when you consider that many of the people funding and implementing the research, are approaching it from a eugenics standpoint. Once again, this is where population reduction plans could potentially be implemented, and targeted to specific groups of people. The availability of this data offers the unprecedented capacity to single out certain genomes or groups for some sort of targeted effects.

We live in the era of big data. At some point, somebody or something (perhaps an artificial intelligence) will figure out how to use this infinite data set for control over the masses, and the reduction of our numbers, if that is in their best interests. The collection of data is the precursory step for setting up and implementing a massive, artificial intelligence total panopticon control grid.

This does admittedly sound like a bad science fiction movie, but it is the stated objective of the transhumanist plan. Merging the minds of men with machines through a brain-computer interface is designed to lead to the inevitable "uploading" of human consciousness into the cloud[108]. This digital immortality is the ultimate goal of the transhuman singularity.

From a philosophical, esoteric, or alchemical point of view, this is the achievement of the "Great Work", the "Philosopher's Stone". Transhumanism is and always has been the ultimate goal of the secret societies and the mystery schools of antiquity.

So, what can be said about the philosophical, esoteric, or alchemical viewpoint of autism, and how does it relate to this? We will explore this avenue of thought in the next chapter...

Autism - Autumn and the Fall of the Mind

Chapter Twenty-Two - Autism: The Alchemical Perspective

Since the origins of transhumanism are derived from the alchemical philosophy of the ancient mystery schools, it is important to understand the alchemical perspective of autism and how it relates to the transhumanist philosophy.

This all may seem to sound like a lot of hocus pocus to many people, but what you need to understand is that, even if you don't believe in any of this stuff, there are people in positions of power who do, and what they do with this knowledge will affect you.

It is imperative to understand this perspective so that you can comprehend why the people who are pushing this agenda do the things that they do. If you have any doubts that people in positions of power are involved with occult or esoteric groups or belief systems, all you need to do is a quick computer search of the terms, "Bohemian Grove", "Cremation of Care Ritual", or "ritual sacrifice at CERN", and there are countless other examples you can find with a very cursory search.

    The point is that there are people within the circles of power who believe and use occult philosophical principles to achieve their goals. So, understanding some of the basic concepts behind this viewpoint will be very helpful in knowing what to expect from those who control the science behind this, and knowing how to move forward.

So, what can we determine about autism from an alchemical standpoint? An important facet in any alchemical or esoteric philosophical study has to do with etymology, or word origins. Language is the instrument of spirit[109], and words have meaning. Sometimes, our modern parlance forgets these meanings, and that is why delving into the origins of words is imperative to garnering a deeper understanding of the ideas being conveyed.

Let's look at the etymological meaning of the word "autism". Derived from the Greek word "autos", it means "self" or "one", and also from the English "ism", which means "isolated" or "separated", so autism means "isolated self", or "separated one". A similar sounding word phonetically is "autumn".

The etymological origin of the word autumn comes from the Latin "auctus", which is also the same root word for August, and it means "increase". It can be argued that the root words for both autism and autumn come from a common source, and thus the esoteric meaning of the term autism would mean "increased self" rather than isolated self. This is the ideology of transhumanism encoded in the word "autism".

This is the reason why autism is being pushed as the next step in human evolution. This is symbolic of an alchemical transmutation of the "fall-en" man into the new man.

This symbolic representation is allegorized as the seasons. Autism is the fall, and the new dawn in the springtime will be the alchemical transmutation of the autist into Human Plus, the transhuman. There are several other levels of meaning that anyone who has studied occult philosophy for some time will be able to single out.

Another example of the alchemical symbolism tying autism to transhumanism is the allegory of the alchemical process itself. When you perform an alchemical feat, there are three basic actions that take place in proper sequence.

The first step in the process is to break something down into its constituite parts. In this example, we would be breaking down human intelligence into its distinctive categories (spatial intelligence, verbal intelligence, emotional intelligence, etc.).

The next phase of the alchemical process would be the purification process. In our example, we would be purifying out the aspects of intelligence that we want to carry forward through the process.

The third and final step in this process is called the alchemical marriage. In our example, this is represented by the merging of the purified aspects of intelligence with machines to achieve the transhuman singularity, the equivalent of the alchemical marriage.

So, to summarize, the concept is to use the autism epidemic to break down human intelligence into its constituite parts, purify out the aspects of intelligence that are desirable for merging with technology, and unifying the man and the machine together into the transhumanist singularity. This is the vision of the transhumanists, and it is a perversion of the natural science known as alchemy. It is a total affront to the natural order, and to the Creator.

The blatant misuse of the autism epidemic to achieve the transhumanist singularity is a total violation of the free will principle, not to mention a gross violation of human rights and dignity. From a philosophical perspective, the transhumanist movement lacks the ethical or moral principles and guidelines that keep our society from imploding.

The transhumanist viewpoint is one of secular humanism and/or moral relativism when it comes to achieving its goals. In the minds of those pushing this transhumanist singularity onto mankind, the ends always justify the means. They will stop at nothing to achieve their "Great Work", to attain the "Philosopher's Stone", to finally reach their ultimate objective from time immemorial.

This achievement that they seek is to become God. It is the great lie told in the garden of Eden. You can be as gods. This is the forbidden fruit that entices those who seek the transhumanist singularity as the answer to their ages old quest.

But, can they truly accomplish this great work? Or is it doomed to failure? The important aspect of alchemical thought that these people are ignoring to achieve these goals, is the principle of karma. You reap what you sow. And they are sowing the seeds of destruction by their actions in their deliberate misuse of the autism epidemic to achieve the transhumanist singularity.

Let's summarize everything that we've learned in this volume, and tie it together into a cohesive hypothesis of how we got to where we are with this epidemic, what to expect next, and where to go from here. Hopefully, by exposing the mechanisms that allow this epidemic to escalate unhindered, we can educate the masses and end the autism epidemic. Let's lay it all out...

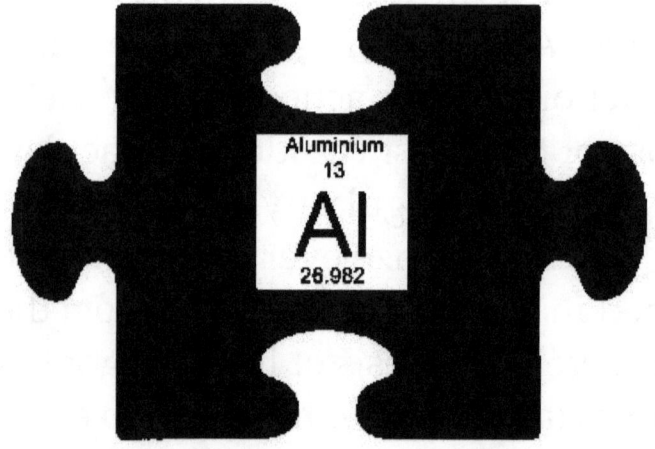

Chapter Twenty-Three - Conclusion - A Theory of Autism as an Engineered Epidemic

The autism epidemic is still growing exponentially. We have explored the history of this disorder, its origins, the detailed study of its characteristics, and the endless litany of probable causes. And, although we understand the general causative factors that contribute to this epidemic, it would seem that nobody will take action and do something, anything, to stop this epidemic in its tracks.

The authority figures in the scientific and medical research communities just keep throwing their hands in the air and saying, "we don't know what causes it, but we do know what doesn't cause it." This is an oxymoron on the face of it.

Autism has been studied intensively for nearly a century now, and rather than figuring out causes, preventative measures, and potential cures, the financiers of "science" have instead capitalized on the "treatment" of autistic symptoms, all the while willfully ignoring the very obvious correlations to certain products and materials.

In fact, cybernetics methodologies applied to the study of this condition have actually figured out its triggering mechanisms, and learned how to manipulate neuronal calcium channels to reproduce these symptoms.

They also developed artificial intelligence machines based upon the perceptual intelligence displayed by autists. The holistic systems approach of the cybernetics studies produced much insight into the condition and its physiological causes.

Endless studies have been done trying to determine what substances and materials may be contributory to the rise in the prevalence of autism. But, although study after study finds significant correlations to certain products and materials (i.e. - vaccines, aluminum, and mercury), the bought and paid for researchers in charge of these studies manipulate and skew the data and/or cover up or destroy the data that does not support the narrative that their corporate sponsors want.

What conclusions could we draw from these facts? First, the information compiled in this book points to one material above all others as the primary factor in the causation of the autism epidemic. That material is aluminum. It keeps showing up all around autism over and over again in the historical timeline in its many varied forms and uses.

So, aluminum is indicated in the onset of autistic symptoms. The researchers that look into this interesting correlation always seem to explain it away and dismiss it. Which leads directly to my second conclusion, there is an active coverup going on with the data on autism and aluminum toxicity.

This means that either the scientists involved in the research are either complacent or complicit. In other words, they either simply look the other way when certain data shows up, or they actively manipulate the data to show something else. They likely have good reasons for going along with the coverup of data, after all, a paycheck is a pretty good motivator.

The third conclusion that I've come to is that the autism epidemic is allowed to continue because it has intrinsic value to the transhumanist movement. It could even be argued that autism is an engineered condition, being tweaked to achieve "autism without intellectual impairment", or a high functioning form of autism like Asperger's Syndrome.

The fourth conclusion I've been able to draw is that those who fall on the autism spectrum are being used as the proverbial canaries in the coalmine, or guinea pigs, to test the waters for new technologies and pharmaceuticals, especially those technologies that relate to the coming transhumanist singularity.

The fifth conclusion I've come to is that the primary delivery systems for autism triggers are vaccines, and also that the powers that be know that vaccines are the primary causative factor leading to autism. This is why the pharmaceutical companies, the news media, and the government health agencies that have a vested interest in vaccines (I'm looking at you, CDC) are pushing so aggressively against what they call "anti-vaxxers".

The sixth conclusion I've come to realize is that this epidemic is being used for mass data collection for the sake of putting everyone in a health database in order to better control every facet of your life. You are a ward of the state, you are property of the small group of family bloodlines at the top of the power structure who seek to become the gods of this world.

The seventh and final conclusion that I've come to is the most important one. Despite what sounds like doom and gloom from the other conclusions I've come to, this conclusion is the one that offers hope.

You see, I concluded that if we, the people, could stand up and put a stop to the vaccine agenda, if we, the people, call out the establishment on the link between autism and vaccines and aluminum, if we the people can end the autism epidemic, then the elitist power structure's whole house of cards comes crumbling down, and their transhumanist control grid will not be able to go into place. Their whole plan to end humanity as we know it falls apart, and we can build a better future for our children.

We must stand together, speak up, and remove our consent from their system, their system that seeks to enslave us, use and abuse us, their eugenics based transhumanist system. If enough people understand and resist their agenda, then we can have a brighter future for all mankind, and not just the select few who think themselves better than the masses. We can have a better future for all, not a future where we are all tied to a computer for eternity in an endless cycle of debt and misery, subject to the whims of the controllers and/or an artificial intelligence control grid.

We have hope. Be strong, be vigilant, be brave, stand up, speak up, and don't back down. Remove your consent from their system. Our children's future depends on it. And remember, God is not mocked.

May God bless you all.

The Abomination of Desolation standing in the holy place...

Your body is the temple of God, and therefore is the holy place...

The Abomination of Desolation is the transhuman singularity...

> (Matt 24:15-16) ¹⁵When ye therefore shall see the abomination of desolation, spoken of by Daniel the prophet, stand in the holy place, (whoso readeth, let him understand:) ¹⁶Then let them which be in Judaea flee into the mountains:

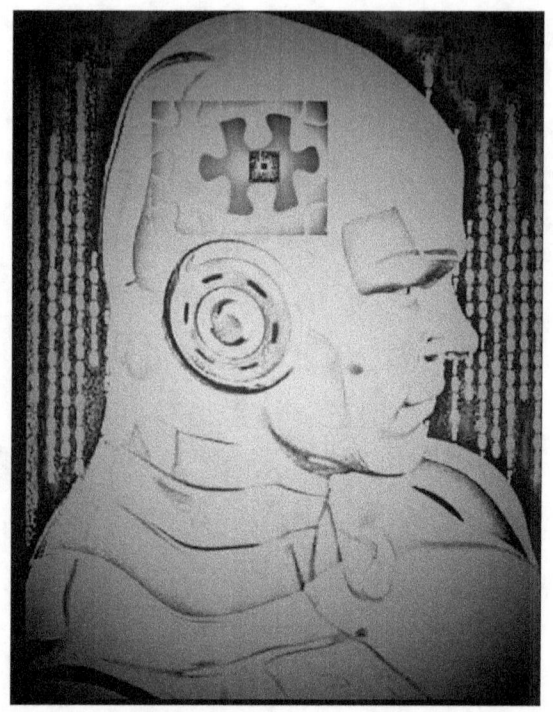

Notes/Sources

1.) www.britannica.com

2.) "Transhumanism and Religion" - Institute for Ethics and Emerging Technologies - John G. Messerly - Jan. 18, 2015

3.) www.immortality-project.org

4.) "Secret of ETERNAL LIFE? We will know what it is by 2029, says Google chief" - Sean Martin - Mar. 20, 2017 - www.express.co.uk

5.) "Postgenderism: Beyond the Gender Binary" - George Dvorsky and James Hughes, Ph.D. - Institute for Ethics and Emerging Technologies - March 2008

6.) "Postmodernity's Search for Postgender: Brophy, Winterson and Place - Univerzita Karlova V Praze - August 2014

7.) www.autismspeaks.com

8.) www.cdc.gov

9.) "Autistic Disturbances of Affective Contact" - Leo Kanner - John's Hopkins University - 1943

10.) "The extreme male brain theory of autism" - Trends in Cognitive Sciences, 6, 248-254

11.) "Sexing the Autistic Brain: Extreme Male?" - Daniel Voyer, Ph.D. - Perceptual Asymmetries - Psychology Today - Oct. 10, 2014

12.) "A Link Between Autism and Testosterone?" - Eben Harrell - Time Magazine - Jan. 15, 2009

13.) "Our Brains Harbor 'Residual Echo' of Neanderthal Genes" - National Institute of Mental Health Section on Integrative Neuroimaging - Michael Gregory, M.D. - July 24, 2017

14.) "Distinct selective forces and Neanderthal introgression shaped genetic diversity at genes involved in neurodevelopmental disorders" - Alessandra Mozzi, Diego Forni, Manuela Sironi - Scientific Reports 7, Article number: 6116 - www.nature.com - 2017

15.) "Did a Drop in Testosterone Civilize Modern Humans?" - Nathan H. Lents, Ph.D. - Beastly Behavior - Psychology Today - Jan. 9, 2017

16.) "New insights into differences in brain organization between Neanderthals and anatomically modern humans " - Eiluned Pearce, Chris Stringer, and R.I.M. Dunbar - Proceedings of the Royal Society B - Biological Sciences - May 7, 2013

17.) New study finds transmen have high autistic symptoms" - www.news-medical.net - Cambridge University - May 6, 2011

18.) "How the only NHS transgender clinic for children 'buried' the fact that 372 of 1069 patients were autistic" - Stephen Adams - Health Correspondent For The Mail On Sunday - www.dailymail.co.uk - Nov. 17, 2018

19.) Definition of cybernetics: the science of communication and control theory that is concerned especially with the comparative study of automatic control systems (such as the nervous system and brain and mechanical-electrical communication systems) - www.merriam-webster.com

20.) "Quantum Processing: The Path from Autistic to Empathic AI?" - Twain Liu - www.startupgrind.com - 2016

21.) "Neanderthal brains re-created in a lab could one day be put into crab-like ROBOTS to create cyborg cavemen, scientists claim" - Aaron Brown for MailOnline - www.dailymail.co.uk - June 29, 2018

22.) "Gender identity and children with autism spectrum disorder" - Dr. Ruth Bevan - The Association for Child and Adolescent Mental Health - July 21, 2017

23.) "There's Growing Evidence For A Link Between Gender Dysphoria And Autism Spectrum Disorders" - Zhana Vrangalova - contributor - Pharma & Healthcare - www.forbes.com

24.) "Fetal testosterone and autistic traits" - Bonnie Auyeng, Simon Baron-Cohen, Emma Ashwin, Rebecca Knickmeyer, Kevin Taylor - British Journal of Psychology 100(1), 1-22, 2009

25.) "The Neanderthal Theory" - Leif Ekblad, Independent Researcher - April 24, 2001 www.rdos.net

26.) "Are there alternative adaptive strategies to human pro-sociality? The role of collaborative morality in the emergence of personality variation and autistic traits" - Penny Spikins, Barry Wright, & Derek Hodgson - Time and Mind > The Journal of Archaeology, Consciousness and Culture - Volume 9, 2016 - Issue 4 - Pages 289-313

27.) "Prevalence of Clinically and Empirically Defined Talents and Strengths in Autism" - Meilleur, A. - A.S., P. Jelenic, and L. Morton. 2015. - Journal of Autism and Developmental Disorders 45(5): 1354-1367 - doi:10.1007/s10803-014-2296-2

28.) "How our autistic ancestors played an important role in human evolution" - Penny Spikins, University of York - March 27, 2017 - www.theconversation.com

29.) "Enhanced mental image mapping in autism" - I. Soulieres, L. Mottron - Neuropsychologia Volume 49, Issue 5, April 2011, Pages 848-857

30.) "Comparing models of intelligence in Project TALENT: The VPR model fits better than the CHC and extended Gf-Gc models" - Jason T. Major, Ian J. Dearly - Intelligence Volume 40, Issue 6, November-December 2012, Pages 543-559

31.) "New insights into differences in brain organization between Neanderthals and anatomically modern humans" - Eiluned Pearce, Chris Stringer, and R.I.M. Dunbar - The Royal Society - Proceedings B - www.ncbi.nlm.nih.gov

32.) "Computer AI passes Turing test in 'world first'" - BBC Technology - June 9, 2014 www.bbc.com

33.) "Why tiny Neanderthal brains are now growing in petri dishes" - Laura Geggel - NBC News Mach - June 27, 2018 - www.nbcnews.com

34.) "History of Cybernetics" - American Society For Cybernetics - www.asc-cybernetics.org

35.) "Organization For Physiological Homeostasis" - Walter B. Cannon - Physiological Reviews Vol. 1 IX, No. 3, July, 1929 - available at www.physiology.org

36.) "Walter Cannon: Homeostasis, the Fight-or-Flight Response, the Sympathoadrenal System, and the Wisdom of the Body" - David Goldstein - Brain Immune Bridging Neuroscience & Immunology - May 16, 2009 - www.brainimmune.com

37.) "Toward a theory of schizophrenia" - Gregory Bateson, Don D. Jackson, Jay Haley, John Weakland - 1956

38.) "The Double Bind Theory: Still Crazy-Making After All These Years" - Paul Gibney - Psychotherapy In Australia - Vol. 12, No. 3 - May 2006

39.) "Autistic Disturbances of Affective Contact" - Leo Kanner - Johns Hopkins University - 1943

40.) "Leo Kanner (1894-1981) Papers Archives Finding Aid" - [Processed by W.E. Baxter, Nov. 12, 1985] - Melvin Sabshin, M.D. Library & Archives - American Psychiatric Association Foundation

41.) "Steps To An Ecology Of Mind" - Collected Essays In Anthropology, Psychiatry, Evolution, And Epistemology" - Gregory Bateson - Jason Aronson Inc. - Copyright 1972

42.) "Autism-like behaviours and germline transmission in transgenic monkeys overexpressing MeCP2" - Zhen Liu, Xiao Li, Zilong Qiu - Nature International Journal of Science 530, 98-102 - Feb. 4, 2016

43.) "The History of Artificial Intelligence" - Rockwell Anyoha - Science in the news - Harvard University - Aug. 28, 2017
www.sitn.harvard.edu

44.) "Machine Intelligence from Cortical Networks (MICrONS)" - www.iarpa.gov/index.php/research-programs/microns

45.) "IARPA Project Targets Hidden Algorithms of the Brain" - Nand Kishor - June 28, 2017 - www.houseofbots.com

46.) "When the world becomes 'too real': a Bayesian explanation of autistic perception" - Elizabeth Pellicano and David Burr - Trends in Cognitive Sciences - TICS-1125 - 2012 http://dx.doi.org/10.1016/j.tics.2012.08.009

47.) "US Bets $100 Million On Machines That Think More Like Humans" - Shelly Fan - March 13, 2016 - Singularity Hub

48.) "Mapping the mind with nanotechnology" - Katherine Sanderson - The Guardian - May 29, 2013

49.) National Alliance of Mental Illness - www.nami.org

50.) "Could Asperger's and Autism Be the Next Step in Evolution?" - Dr. Frank Gaskill - Nov. 3, 2017 - www.psychbytes.com

51.) "Questions for Richard Tsien: Taking apart autism's machinery" - Ann Griswold - March 22, 2016 - www.spectrumnews.com

52.) "Unifying Views of Autism Spectrum Disorders: A Consideration of Autoregulatory Feedback Loops" - Mullins C., et. al. - Neuron 2016 - PubMed - www.ncbi.nlm.nih.gov

53.) "Molecular mechanisms of autism: a possible role for Ca^{2+} signaling" - Jocelyn F Krey, Ricardo E Dolmetsch - current opinion in neurobiology 17(1), 112-119, 2007-Feb. - www.sciencedirect.com

54.) "Disruption of neuronal calcium homeostasis after chronic aluminum toxicity in rats." - Amarpreet Kaur, Kiran DipGill - Basic & Clinical Pharmacology & Toxicology Volume 96 Issue 2 - www.ncbi.nlm.nih.gov

55.) "Aluminum-induced oxidative DNA damage recognition and cell-cycle disruption in different regions of rat brain" - Kumar V, et.al. - Toxicology 2009 - PubMed - www.ncbi.nlm.nih.gov

56.) "Regional alterations in calcium homeostasis in the primate brain following chronic aluminum exposure." - Savin S, Julka D, Gill KD - Mol Cell Biochem. 1997 - www.ncbi.nlm.nih.gov

57.) "Altered calcium homeostasis: a possible mechanisms of aluminum-induced neurotoxicity." - Julka, D, Gill KD - Biochem Biophys Acta. - 1996 - www.ncbi.nlm.nih.gov

58.) "Autism and EMF? Plausibility of a pathophysiological link - Part 1." - Herbert MR, et.al. - Pathophysiology 2013 - www.ncbi.nlm.nih.gov

59.) "Scientific evidence contradicts findings and assumptions of Canadian Safety Panel: microwaves act through voltage-gated calcium channel activation to induce biological impacts at non-thermal levels, supporting a paradigm shift for microwave/ lower frequency electromagnetic field action." - Pall ML. - Rev Environ Health 2015 - www.ncbi.nlm.nih.gov

60.) "Stratospheric Welsbach seeding for reduction of global warming" - U.S. Patent # US5003186A

61.) "Elevated silver, barium, and strontium in antlers, vegetation, and soils sourced from CWD cluster areas: do Ag/Ba/Sr piezoelectric crystals represent the transmissible pathogenic agent in TSE's?" - Purdey M - Med Hypotheses 2004;63(2): 211-25, PubMed - www.ncbi.nlm.nih.gov

62.) "Are there alternative adaptive strategies to human pro-sociality? The role of collaborative morality in the emergence of personality variation and autistic traits" - Penny Spikins, Barry Wright, & Derek Hodgson - Time and Mind > The Journal of Archaeology, Consciousness, and Culture - Volume 9, Issue 4, 2016

63.) www.autismacceptancemonth.com

64.) "An Introduction to the Overton Window of Political Possibilities" - Nathan J. Russell - Jan. 4, 2006 - Mackinac Center For Public Policy - www.mackinac.org

65.) "Autism and human evolutionary success" - www.sciencedaily.com - November 15, 2016

66.) "Russia 2045 Avatar Project" - www.artificialbrains.com

67.) "Brain-computer interface game applications for combined neurofeedback and biofeedback treatment for children on the autism spectrum" - Elisabeth V.C. Friedrich, Neil Suttie, Aparajithan Sivanathan, Theodore Lim, Sandy Louchart, and Jaime A. Pineda - Frontiers in Neuroengineering - July 3, 2014 - https://doi.org/10.3389/fneng.2014.00021

68.) "Culling the herd: eugenics and the conservation movement in the United States, 1900-1940" - Allen GE - J Hist Buol. 2013 Spring; 46(1): 31-72 - PubMed - www.ncbi.nlm.nih.gov

69.) www.americanrhetoric.com

70.) "A Rhetorical Model of Autism: a Pop Culture Personification of Masculinity in Crisis" - Dissertation - Malcolm Matthews, Doctor of Philosophy - Interdisciplinary Humanities Ph.D. Program - Brock University - 2017

71.) "How our autistic ancestors helped shape human evolution" - Penny Spikins - Independent - Dec. 1, 2017 - Genetic Literacy Project: Science Not Ideology - www.geneticliteracyproject.org

72.) "Propaganda" - Edward Bernays - 1928

73.) "History's 30 Most Inspiring People On The Autism Spectrum" - Applied Behavior Analysis Programs Guide - www.appliedbehavioranalysisprograms.com

74.) www.britannica.com

75.) "The Triangle of Enhancement Medicine, Disabled People, and the Concept of Health: A New Challenge for HTA, Health Research, and Health Policy " - Alberta Heritage Foundation For Medical Research - HTA Initiative # 23 - Dec. 2005

76.) "5 Neuroscience Experts Weigh in on Elon Musk's Mysterious 'Neural Lace' Company" - Eliza Strickland - April 12, 2017 - IEEE Spectrum - https://spectrum.ieee.org

77.) "Brain Augmentation Could Help People Live Better Lives by 2030" - Peter Hess - July 4, 2017 - Inverse - www.inverse.com

78.) "Super Soldiers: The quest for the ultimate human killing machine" - Michael Hanlon - Nov. 17, 2011 - www.independent.co.uk

79.) "Savant-like Numerosity Skills Revealed in Normal People By Magnetic Pulses" - Allan Snyder, Homayoun Bahramali, Tobias Hawker, D. John Mitchell - Sage Journals Vol. 35, Issue 6, Pages 837-845 - June 1, 2006

80.) "Military Applications of Invasive Brain Stimulation" - Melanie Segado - June 29, 2017 - IEEE Technology and Society https://technologyandsociety.org

81.) "The Military Is Seeking Out People With Autism - Here's Why" - The Autism Site - blog.theautismsite.com

82.) "Neurodiversity & Cybersecurity Careers: Recruiting & Retaining Autistic Cybersecurity Professionals" - Eleanor Dallaway - March 8, 2017 - InfoSecurity Magazine - www.infosecurity-magazine.com

83.) "What Does the Word 'Autism' Mean?" - www.webmd.com

84.) "The History of Aluminum" - www.aluminiumleader.com

85.) "The Environmental Literacy Council - www.enviroliteracy.org

86.) "What is Nanotechnology" - www.nano.gov

87.) "What Is CRISPR?" - Aparna Vidyasagar - April 20, 2018 - www.livescience.com

88.) "CRISPR-CAS9: GENE DRIVE" - Wyss Institute Technologies - https://wyss.harvard.edu

89.) "Hidden Risk un Foods and Cosmetics" - Jan. 20, 2016 - www.mercola.com

90.) "Advances in aluminum hydroxide-based adjuvant research and its mechanism" - Peng He, Yening Zou, and Zhongyu Hu - Human Vaccines & Immunotherapeutics - Taylor & Francis - Feb. 11, 2015 - PubMed - www.ncbi.nlm.nih.gov

91.) "Phospholipid Bilayer-Coated Aluminum Nanoparticles as an Effective Vaccine Adjuvant-Delivery System" - Ting Wang, Yuanyuan Zhen, Xiaotu Ma, Biao Wei, and Ning Wang - ACS Applied Materials & Interfaces Vol. 7, Issue 12: Pages 6391-6396 - March 17, 2015

92.) "Nanotechnology-inspired Grand Challenges in the United States" - Mike Roco - NSF and NNI - US-Korea Nano Forum, Seoul, September 26, 2016

93.) "DTFN Estimates for Nano Domestic Quell Phase 4 Updated Compliance" - Department of Defense, June, 2013 -   leaked document

94.) "Future Strategic Issues/Future Warfare [Circa 2025] - Dennis M. Bushnell, Chief Scientist - NASA Langley Research Center - July 2001

95.) "The 6 Most Evil Human Experiments Perpetrated By The U.S. Government" - Richard Stockton - Oct. 24, 2017 - https://allthatsinteresting.com

96.) "Ed Boyden: The brain is like a computer, and we can fix it with nanorobots" - Interviewed by Ian Tucker - March 11, 2012 - The Guardian

97.) "Biosynthesis of nanoparticles: technological concepts and future applications" - Pra Shant Mohanpuria, Nisha K. Rana, Sudesh Kumar Yadav - J Nanopart Res (2008) 10:507-517 - Springer Science + Business Media - Aug. 3, 2007

98.) "Medical Nanobots Will Connect Brain To Cloud Computing - Ray Kurzweil" - www.thesleuthjournal.com

99.) "Nanowire nanocomputer as a finite-state machine" - Jun Yao, Hao Yan, Shamik Das, James F. Klemic, James C. Ellenbogen, and Charles M. Lieber - Proceedings of National Academy of Sciences of the United States if America - Jan. 27, 2014

100.) "Kurzweil Claims That The Singularity Will Happen By 2045" - published by Chriatianna Reedy in Future Society - www.futurism.com

101.) "Heads in the cloud: Scientists predict internet of thoughts 'within decades'" - Neuroscience News - April 12, 2019 - www.neurosciencenews.com

102.) "EU 5G Appeal - Scientists warn of potential serious health effects of 5G" - www.jrseco.com

103.) "Raytheon Delivers Non-lethal Sheriff Active Denial System" - www.microwavejournal.com - Nov. 1, 2005

104.) "Our Cyborg Future: Law and Policy Implications" - Benjamin Wittes and Jane Chong - September 2014 - Brookings Institute - www.brookings.edu

105.) NIMH Data Archive - https://nda.nih.gov

106.) Global Alliance for Genomics & Health - www.ga4gh.org

107.) "What is blockchain?" - www.bankrate.com

108.) "Virtual Brain Using Nanobots" - Sathya Priya M, Gomathi S - International Journal of Emerging Technology & Research - Vol. 1, Issue 1, Nov.-Dec., 2013 - www.ijetr.org

109.) "The Mystery of the Cathedrals" - Fulcanelli - 1926

www.ingramcontent.com/pod-product-compliance
Lightning Source LLC
Chambersburg PA
CBHW072133170526
45158CB00004BA/1352